W9-BZL-620

Erwin Engeler

Foundations
of Mathematics

Questions of Analysis,
Geometry & Algorithmics

Translated by Charles B. Thomas
With 29 Figures

Springer-Verlag
Berlin Heidelberg New York
London Paris Tokyo
Hong Kong Barcelona
Budapest

Author:
Erwin Engeler
Mathematikdepartment
ETH-Zentrum
CH-8092 Zürich, Switzerland

Translator:
Charles B. Thomas
DPMMS, 16 Mill Lane
Cambridge CB2 1SB, Great Britain

Title of the original German edition:
Metamathematik der Elementarmathematik in the Series Hochschultext
© 1983 by Springer-Verlag
ISBN 3-540-12151-X

Mathematics Subject Classification (1991): 03-XX

ISBN 3-540-56422-5 Springer-Verlag Berlin Heidelberg New York
ISBN 0-387-56422-5 Springer-Verlag New York Berlin Heidelberg

Library of Congress Cataloging-in-Publication Data
Engeler, Erwin. [Metamathematik der Elementarmathematik. English] Foundations of mathematics :
questions of analysis, geometry & algorithmics / Erwin Engeler ; translated by Charles B. Thomas. p. cm.
Includes bibliographical references.
ISBN 0-387-56422-5
1. Metamathematics. I. Title. QA9.8.E5413 1993 510'.1–dc20 92-46107

© Springer-Verlag Berlin Heidelberg 1993
Printed in Germany

Typesetting: Camera-ready by author/translator using Springer TEX macropackage
41/3140 - 5 4 3 2 1 0 – Printed on acid-free paper

Preface

This book appeared about ten years ago in German. It started as notes for a course which I gave intermittently at the ETH over a number of years. Following repeated suggestions, this English translation was commissioned by Springer; they were most fortunate in finding translators whose mathematical stature, grasp of the language and unselfish dedication to the essentially thankless task of rendering the text comprehensible in a second language, both impresses and shames me. Therefore, my thanks go to Dr. Roberto Minio, now Darmstadt and Professor Charles Thomas, Cambridge. The task of preparing a LaTeX-version of the text was extremely daunting, owing to the complexity and diversity of the symbolisms inherent in the various parts of the book. Here, my warm thanks go to Barbara Aquilino of the Mathematics Department of the ETH, who spent tedious but exacting hours in front of her Olivetti.

The present book is not primarily intended to teach logic and axiomatics as such, nor is it a complete survey of what was once called "elementary mathematics from a higher standpoint". Rather, its goal is to awaken a certain critical attitude in the student and to help give this attitude some solid foundation. Our mathematics students, having been drilled for years in high-school and college, and having studied the immense edifice of analysis, regrettably come away convinced that they understand the concepts of real numbers, Euclidean space, and algorithm.

The advocacy of a critical attitude towards time-honoured teachings does not mean a denial of the usefulness, let alone the necessity, of a general consensus in mathematics. What I vehemently oppose is the lack of imagination that underlies the tendency to blind acceptance and thoughtless compliance with which many students approach the basic attitudes and concepts of mathematics. To combat this tendency, mathematical logic has created technical machinery, in terms of which such a critique can be cogently, in fact mathematically, expressed. The word "metamathematics" in the original German title of this book thus indicates the method; "elementary mathematics" indicates the subject area of the critique, namely, analysis, geometry and algorithmics.

In a nutshell: how does one arrive at the axioms of elementary mathematics and what do we gain by having them?

Zürich, September 1992   E. Engeler

Contents

Chapter I. The Continuum

§1 What Are the Real Numbers?

About one hundred years ago, *Dedekind* was Professor of Mathematics at the ETH in Zurich, teaching differential and integral calculus. He describes how teaching this particular subject he was confronted with the questions of the foundations of analysis. Dedekind's little book *"Was sind und was sollen die Zahlen"*[1], which still makes enjoyable reading, reflects his ingenious approach to the problem.

Dedekind's solution is now well-known. It consists of a program for the *arithmetization* of the continuum – the reduction of the basic ideas of analysis to notions involving natural numbers. In this way one should obtain unassailable proof of the existence of the mathematical structure underpinning analysis: all the properties of the continuum that are used in analysis ought to arise from this construction.

The intuition underlying the construction of $\mathbb{Q}$ is of course the representation of integral fractions $\frac{a-b}{c}$ as triples $\langle a, b, c \rangle$ together with the corresponding presentation of the manipulative rules for addition, multiplication, order. The construction of $\mathbb{R}$ rests on the intuitive notion of completeness, which we shall discuss in Section 2 below. Essentially the idea is that for each change of sign of each continuous function there should be a real number at which that function's value is zero. Thus the notion of number is dependent upon that of function. This is already implicit in the title of Dedekind's paper "Stetigkeit und irrationale Zahlen[2]", (1872).

The main objection to Dedekind's program is that it is not pure: the construction does not just use the natural numbers and the operations and relations such as addition, multiplication, and order defined on them, but also uses objects and concepts from a "higher" region, such as sets (of natural numbers). The construction itself is not arithmetic, but set–theoretic. But surely set theory is further removed from intuition than the continuum itself, which we can (almost) imagine. So it can reasonably be argued that Dedekind's program has not answered the question posed in the title of this section.

---

[1] Numbers: what they are and what they are for.
[2] Continuity and irrational numbers.

Dedekind's Arithmetization of the Continuum (Sketch)

(a) $N = \langle \mathbb{N}, 0, 1, +, \cdot, \leq \rangle$ satisfies Peano's axioms:

   (i) $0 \neq x + 1$ for all $x \in \mathbb{N}$.

   (ii) $x + 1 = y + 1$ only if $x = y$, for all $x, y \in \mathbb{N}$.

   (iii) If $M \subseteq \mathbb{N}$ and $M \neq \emptyset$ then M has a smallest element with respect to $\leq$.

   (iv) $x \leq y$ precisely when there is a $z \in \mathbb{N}$ such that $x + z = y$.

   (v) Addition and multiplication satisfy the recursive equations

$$x + (y + 1) = (x + y) + 1 \qquad x + 0 = x$$
$$x \cdot (y + 1) = x \cdot y + x \qquad x \cdot 0 = 0$$

(b) $\mathcal{Q} = \langle \mathbb{Q}, 0, 1, +, \cdot \ \leq \rangle$ is an ordered field, where $\mathbb{Q}$ itself is the set of equivalence classes of triples $\langle a, b, c \rangle \in \mathbb{N}^3$ (with $c \neq 0$) under the equivalence relation defined by $\langle a, b, c \rangle \equiv \langle a', b', c' \rangle$ iff $ac' + b'c = a'c + bc'$. Addition is defined by $\langle a, b, c \rangle + \langle a', b', c' \rangle :=$ $\langle ac' + a'c, bc' + b'c, cc' \rangle$, and multiplication by $\langle a, b, c \rangle \cdot \langle a', b', c' \rangle :=$ $\langle aa' + bb', ab' + ba', cc' \rangle$. Zero is the equivalence class of $\langle 0, 0, 1 \rangle$; One is the equivalence class of $\langle 1, 0, 1 \rangle$, the ordering is given by $\langle a, b, c \rangle \leq \langle a', b', c' \rangle$ iff $ac' + b'c \leq a'c + bc'$.

(c) $\mathcal{R} = \langle \mathbb{R}, 0, 1 +, \cdot, \leq \rangle$ is a complete ordered field; $\mathbb{R}$ consists of Dedekind cuts, i.e., nonempty subsets $S \subseteq \mathbb{Q}$, such that $S \neq \mathbb{Q}$, S has no greatest element, and whenever $x \in S$ and $y \leq x$ then $y \in S$.

   Addition $\qquad\qquad$ $S + T = \{x + y \mid x \in S, y \in T\}$.

   Zero $\qquad\qquad\qquad$ $0 = \{x \in \mathbb{Q} \mid x < 0\}$.

   One $\qquad\qquad\qquad$ $1 = \{x \in \mathbb{Q} \mid x < 1\}$.

   Relation of order $\qquad$ $S \leq T$ iff $S \subseteq T$.

One would like to define multiplication as

$$S \cdot T = \{x \cdot y \mid x \in S, \quad y \in T\},$$

which leads however to (superable) difficulties (practice exercise).

A mathematician may take the view that it is of no concern to him from where the real numbers come – what interests him are the properties of the real numbers. In other words, what he expects from the study of the foundations of analysis is an axiomatization, a listing of properties of the real numbers from which all the theorems of analysis follow by pure logic.

At the turn of the century, Hilbert produced just such an axiomatization. The real numbers are characterized as a complete ordered field; that is, an ordered field in which every ordered set has a least upper bound. This is all one needs because one can show that, up to isomorphism, there is only one complete ordered field. Technically, the situation is described by saying that axiomatization is categorical. We postpone proving this fact until later. (In point of fact Hilbert's original axiomatization differs from the one sketched above, but they are equivalent.) Hilbert characterized the real numbers as an ordered field which is Archimedean and maximal for this property. (To say that it is Archimedean is to require that any positive element can be added to itself a sufficient number of times so as to exceed any given positive element. It is maximal in the sense that any extension field will fail to be Archimedean.) In fact it was just the analysis of the proof of categoricity which led to the final formulation of the axioms.

Even this axiomatic approach leaves much to be desired: essentially the same objections can be raised as against the so-called arithmetization. The problem is that the axioms do not refer only to the basic algebraic operations and ordering relations, $+, \cdot, \leq$, etc, and to the elements of the field to be characterized, but once again to higher–level concepts — "bounded sets" and the like. Consider for example the Archimedean axiom: "Given $a, b > 0$, there exists a positive integer n such that $a \cdot n > b$". One could argue that the theory of natural numbers can be presupposed in any axiomatization. But, in that case, we ought to insist that the axioms for the natural numbers be omitted from the axiomatization as a matter of convention only – that we consider them to be tacitly included in axiomatization.

We ought therefore to add a version of Peano's axioms, and in particular: "Every set of natural numbers which contains 0 and which contains $n + 1$ for every element n is the whole set itself." And again we are speaking of sets. And we do this without having axiomatized the framework in which we employ set–theoretic notions. Now, one could try to axiomatize just enough set theory – namely just those parts of set theory that we need in this connection and once and for all make the convention, that these axioms be tacitly understood as part of every axiomatization.

What does this convention really mean? Are the axiomatic foundations of set theory themselves free of controversy? Certainly not (more about this later), so we must retreat to a more modest position. In order to build such a refuge, in the following sections we shall make use of the technique of formalizing mathematics.

It is no surprise, says the constructivist, that you have difficulties in providing analysis with a set-theoretic foundation. Actual infinites are not part of our intuition; we have no direct perception of their properties.

- Infinite sets could be thought of as a kind of Platonic idea common to all mankind, we might have experienced them in a paradise from which we have been unfortunately exiled. This is muddled ontology. So considered the continuum would fall into the same class of objects as the unicorn or the Easter bunny (we can agree about some of their characteristics, but are less than certain about others).

- Axiomatic set theory may suffice technically as the ultimate basis of mathematics, but it tends to side-step the foundational problems, and the paradoxes in particular, rather than explain them. What's more, the evasion is rather crude: it is not even plausible that the sole source of difficulties lies in the size of the "paradoxical" sets.

- Even the ordinary deductive methods of mathematics are not to be trusted; the Law of the Excluded Middle applied to infinite totalities is particularly suspect. Classical logical reasoning only makes sense if mathematical statements are interpreted as "That is so and so ..." about something which exists; it is questionable whether it is at all meaningful to make classical inferences about infinite entities. Rather, a mathematical statement should be taken to mean: "I have a construction, which is a proof that ...".

How should one react to this attack on classical mathematics?

1. The non–answer goes something like this. You're right, infinite sets don't really exist, and my logic may be shaky. But I don't mind; I can prove results, and I can make a living from the applications. So let me just carry on and "pretend". In any case, questions about existence and construct ability belong to the realm of applied mathematics – and are surely well provided for there? In the worst case I can always hide behind the attitude that mathematics is just a formal, symbolic activity.

2. The majority of mathematicians respond with the following pragmatic answer.

   - The doubts expressed about classical mathematics seem to be even less concrete than classical mathematics itself. The usual conclusions about infinity are quite convincing to the intelligent mathematician. In fact, attempts to eliminate infinite notions wherever possible do not reinforce our conviction about theorems. If anything, experience shows that finitistic, formalized proofs tend to lead to computational confusions, which are hard to discover; conceptual proofs, even if they use infinitary notions, provide more insight and are more convincing.

- What's more, if a contradiction were to appear somewhere, the discovery would be no disaster, but on the contrary interesting and fruitful for classical mathematics.

- Finally, all these various theories about the foundations of mathematics can in fact be discussed within classical mathematics itself; this universality which makes it particularly attractive. Until something better achieves general acceptance, therefore, let us sensibly leave things as they are.

3. The encouraging answer: I can well imagine that someone is startled by the questions thrown up. But it seems to me that they should serve as signposts for further research.

4. Counterattack is also a reply. Without doubt, as has already been pointed out, every precise formulation of the constructive point of view also has its weaknesses.

There is little point in elaborating these topics further now. We should emphasize that our presentation of the various standpoints has been extremely superficial. To rectify this we postpone the discussion until we have sufficient material at our disposal. For the moment let me just mention a few sources from the literature on foundations, for further reading.

Further Reading

Dedekind, R.: Stetigkeit und irrationale Zahlen, 1872, in: R. Friche, E. Noether, O. Ore: Dedekind gesammelte mathematische Werke, vol. 3, pp. 315-334. Braunschweig, Vieweg, 1932

Dedekind, R.: Was sind und was sollen die Zahlen? 1887, in: R. Friche, E. Noether, O. Ore: Dedekind gesammelte mathematische Werke, vol. 3, pp. 335-391. Braunschweig, Vieweg, 1932

Hilbert, D.: Ueber den Zahlbegriff, zu finden im Anhang VI der Grundlagen der Geometrie, 7. Auflage, Stuttgart, Teubner, 1930

Hilbert, D.: Ueber das Unendliche, Mathematische Annalen, vol. 95, pp. 161-190, (1926)

Brouwer, L.E.J.: Begründung der Funktionenlehre unabhängig vom logischen Satz vom ausgeschlossenen Dritten, 1923, in: A. Heyting: L.E.J. Brouwer collected works, vol. 1, pp. 246-267. Amsterdam, North-Holland, 1975

Brouwer, L.E.J.: Zur Begründung der intuitionistischen Mathematik, I, II & III, 1925-1927, in: A. Heyting: L.E.J. Brouwer collected works, vol. 1, pp. 301-314, 321-340, 352-389. Amsterdam, North-Holland, 1975

Bernays, P.: Sur le Platonisme dans les Mathématiques, L'Enseignement Mathématiques 34 (1935), English translation in P. Benacerraf & H. Putnam: Philosophy of Mathematics, selected readings, pp. 274-286, Englewood Cliffs, Prentice–Hall, 1964

Weyl, H.: Ueber die neue Grundlagenkrise der Mathematik, Mathematische Zeitschrift, 10. pp. 39-79, (1921), also in Selecta Hermann Weyl. pp. 211-248, Basel & Stuttgart, Birkhäuser Verlag, 1956

§ 2 Language as Part of Mathematics

Modern logic owes its existence to a truly grandiose dream – one already dreamed by Leibniz. Before recounting the dream, let me describe the historical context.

Leibniz lived at a time when the victorious advance of modern mathematical notation, particularly in algebra and analysis, had begun. Leibniz himself contributed significantly to this advance, for example by introducing the notation we now use for derivatives and integrals. Moreover, he was fully aware of the importance of this development: modern mathematics owes its unparalleled success above all to abstraction. Notation relieves mathematicians from the need always to think of the content of mathematical signs and allows them instead literally to compute with the abstractions themselves. Also in Leibniz's time the axiomatic geometry of the ancient Greeks had undergone a revival and was flourishing once again. The paradigm, Axiom – Theorem – Proof – Definition – Theorem – Proof – ..., influenced wide areas of philosophy. (For example, consider Baruch Spinoza's Ethica, ordine geometrico demonstrata, Ethics, treated in the manner of geometry.)

Shouldn't it be possible, thought Leibniz, to formulate the rules of mathematical proof in such a way that, in applying them, one no longer has to pay attention to the meanings of the expressions being manipulated? What is needed is a calculus ratiocinator: a calculus in which natural reasoning is replaced by formal calculation, so that proof itself becomes an object of mathematics. A prerequisite for such a calculus is the availability of a symbolism in which the axioms, theorems, and definitions of mathematics can be expressed. It was Leibniz's aim to provide just such a symbolism with his formal language, the famous characteristica universalis. Even for a person of Leibniz's genius, however, the time was not ripe for the development of modern logic; so it is unfortunate, but perhaps inevitable, that he did not complete this work. The formal language which was to provide a framework for solving all mathematical questions by calculation (according to Leibniz's motto calculemus) remained a dream.

It was not until the twentieth century that the most important steps were taken in the direction pointed out by Leibniz. Historically, the situation was similar to that in Leibniz's time. There were the works of Dedekind and Cantor, which aimed to reduce the whole of mathematics to set theory, those of Boole, Peano, Peirce, Schröder, which introduced rudimentary mathematical symbolism for the laws of thought. These works inspired boundless optimism in the power of formalization. In the shape of modern logical calculi formalization attained a higher level of rigor at the hands of first-rate mathematicians like

Frege, Russell, Whitehead, Hilbert, Bernays, Gödel, and Church. As a result, it is now possible to talk of language as a part of mathematics, and to discuss the possibility and practicability of a realization of Leibniz's dream.

The Symbolism of Modern Logic

Propositional Logic deals with elementary propositions that are not further analyzed. Propositions may be combined by means of the basic connectives $\wedge, \vee, \neg, \supset, \equiv$ to form compound propositions.

The primitives of the *Logic of Classes* are propositions about properties: $A(x)$, the object x has the property A. (The totality of x's with the property A could itself be treated as an entity, namely the class A. In this case operations performed on the propositions by using the connectives correspond to the so–called Boolean operations on the corresponding classes.)

The *Theory of Relations* adds assertions about relations between objects to these primitive propositions: $A(x,y), B(x,y,z), \dots$; although the language is formal, the intended meaning is often revealed by the use of customary mathematical notation such as $x \le y, x \in y, \dots$.

Predicate Calculus introduces in addition the quantification of assertions with the notation $\exists x, \forall y$.

For now this concludes our description of the symbolism. The reader is expected to be familiar with the notations and their meanings, and to a lesser extent also with the main results of Predicate Calculus.

The optimism about formalization mentioned above stimulated ingenious and talented mathematicians to achievements involving great efforts and persistence. Consider, for example, Principia Mathematica by Whitehead and Russell (1910-1913), Peano's Formulaire de mathématique (1894-1908), or, even more to the point, the work which came first and set the example for others to follow, Frege's Begriffsschrift (1897) and Grundgesetze der Arithmetik (1893- 1903). In his Grundgesetze Frege attempted the reduction of the whole of mathematics to logic. This appeared to be possible through the identification of properties with concept extensions. Extensions of concepts were among the objects of Frege's universe. Instead of saying a is a B, or $B(a)$ for short, one thinks of the extension of the property B as an object b and writes a is in b, or formally $a \in b$. Which extensions are admitted as objects is determined by the choice of formal language, in this case the language of predicate logic with no non–logical predicates except the notion ... is a ..., i.e. the binary predicate symbol $\in$. Equality is not a primitive of the language; it is defined; objects are equal if they are indistinguishable — if they belong to precisely the same extensions. Thus x is equal to y precisely when x is in those z's to which y belongs. The formalism of Frege's Begriffsschrift is rather involved, so we shall write down his system using modern notation. In June 1902 Russell wrote to Frege pointing out that his system was inconsistent (a correspondence well worth reading). The reduction of mathematics to logic through the identification of the notions of set and property was thus unsuccessful.

Frege's Calculus for Pure Thought   (in modern notation)

Language and Logic: Predicate Calculus without equality; variables $x, y, z, \ldots$; predicate symbol $\in$.

Definition: $x = y \equiv_{df} \forall z(x \in z \equiv y \in z)$

Axioms:

(i) Extensionality

$$\forall x \forall y (\forall z(z \in x \equiv z \in y) \supset x = y)$$

(ii) Class construction

$$\exists y \forall x(x \in y \equiv A(x)),$$

for any formula $A(x)$ with precisely the free variable x.

These axioms legitimize the notation $\{x \mid A(x)\}$.

Russell's Paradox: Consider the object $R = \{x \mid x \notin x\}$; neither $R \in R$ nor $R \notin R$ can hold – a contradiction.

The problems encountered by Frege's approach dampened foundational research. Zermelo, a professor in Zurich, was the first to produce a satisfactory axiomatisation. Supplemented by the Axiom of Replacement, Zermelo's axiomatization is one of the two most widely accepted logical bases of mathematics. The other, by Bernays, is also a product of Zurich.

Let us return to the main topic: what are the real numbers? We will present two axiomatizations; one is developed with algebra in mind, the other has more the flavor of analysis. Both axiomatizations use first-order predicate calculus. The formal language will no longer be used, as it was by Frege, to generate the universe, just to describe it. That is to say, the structure is thought of as being given, and the aim is to formulate its characteristic properties. The real numbers are characterized as a complete ordered field.

Elementary Theory of the Real Numbers

$$\mathcal{R} = \langle \mathbb{R}, \leq, +, \cdot, -, {}^{-1}, 0, 1 \rangle.$$

Language and logic: First–order predicate calculus with equality.

Individual variables :	$x, y, z, \ldots$.
Individual constants :	$0, 1$.
Function symbols :	$+, \cdot$ (binary); $-, {}^{-1}$ (unary).
Basic predicate :	$\leq$ (binary).

Axioms:

  (i) Axioms for fields.

 (ii) Order axioms:

$$x \leq x, x \leq y \wedge y \leq x . \supset x = y,$$
$$x \leq y \wedge y \leq z . \supset x \leq z, x \leq y \vee y \leq x,$$
$$x \leq y \supset x + z \leq y + z$$
$$z > 0 \wedge x < y . \supset x \cdot z \leq y \cdot z.$$

(iii) Completeness axiom: from any formula $A(x)$ with precisely one free variable x, we have

$$\exists x A(x) \wedge \exists b \forall x (A(x) \supset x \leq b).$$
$$\supset \exists b [\forall x (A(x) \supset x \leq b) \wedge \forall c (\forall x (A(x) \supset x \leq c) \supset b \leq c)].$$

The axioms for fields and the order axioms require no further explanation, but there is more to say about the completeness axiom. First of all, the notion of completeness has played an especially interesting role in the history of mathematical ideas. Secondly, the question arises, to what extent does the formalization we have given capture the "full content" of the original set–theoretic completeness axiom. Let us examine this question first.

The available formal language does not provide any way to treat sets as individuals in the universe. In order to be able to deal with sets we borrow an idea of Frege's and introduce sets metatheoretically as extensions of predicates (but without taking the dubious step of giving these extensionally described totalities the status of individuals). Instead of talking about sets, we talk about the properties by which they are defined — in particular, those properties of individuals (real numbers) which can be expressed in the given language. For example, the assertion that the set of real numbers x with the property $A(x)$

is non–empty is expressed simply as $\exists x A(x)$; the existence of an upper bound can be formulated by $\exists b \forall x (A(x) \supset x \leq b)$, and so on. This is what enables us to formulate the completeness axiom as above. The question of the extent to which this axiom falls short of the "full" set–theoretic completeness axiom is now really just the question of what class of sets can be defined by properties $A(x)$. The answer will be given in Section 3.

No summary can be a substitute for a study of the primary literature, and the following remarks on the history of the notion of completeness are no exception. (Some excellent editions of original writings are given in the references.) In Greek mathematics the concept of number was closely tied to measurement. The rational numbers served this purpose, satisfying the needs of the philosophers, particularly the Pythagoreans. The discovery of the irrationality of $\sqrt{2}$ and the problem of squaring the circle prompted the development of a mathematics of measurable magnitudes, one of the most magnificent and strikingly modern creations of antiquity. Developed by Eudoxus, it is to be found in the Fifth Book of Euclid. We present it here, in modern guise, by characterizing an Archimedean system of magnitudes as a linearly ordered Abelian semigroup satisfying the Archimedean axiom.

Archimedean Systems of Magnitudes

Intended Structure: $\boldsymbol{G} = \langle G, \leq, + \rangle$

Definition: For $n \in \{1, 2, 3, \ldots\}$ let

$$n \cdot a =_{df} \underbrace{a + a + \ldots + a}_{n \text{ times}}.$$

Axioms:

(i) Abelian semigroup with relative complement:

$$x + y = y + x, \quad (x + y) + z = x + (y + z),$$
$$x + y \neq x, \quad x \neq y \supset . \exists z (x + z = y) \vee \exists z (x = y + z).$$

(ii) Order axioms:

$$x \leq x, x \leq y \wedge y \leq x . \supset x = y,$$
$$x \leq y \wedge y \leq z . \supset x \leq z, x \leq y \vee y \leq x,$$
$$x \leq y \supset x + z \leq y + z.$$

(iii) Archimedean axiom:

$$\forall x \forall y \exists n (x \leq n \cdot y).$$

The positive integers, the positive rationals and the positive real numbers clearly form Archimedean systems of magnitudes. The restriction to positive numbers is reasonable: after all the numbers are for measuring magnitudes. We shall remark on the "impure" form of the Archimedean axiom later. Eudoxus/Euclid show next how the ordering and addition defined in a system of magnitudes can be extended in a straightforward way to (formal) proportions $(x : y)$. With the extended definitions they prove a theorem which would nowadays be expressed as follows:

Every Archimedean system of magnitudes can be isomorphically embedded in the positive real numbers.

(This theorem provides a good opportunity to sketch the promised proof of the categoricity of complete ordered fields by showing that every complete ordered field also satisfies the Archimedean axiom, and is therefore isomorphic to a subfield of $\mathcal{R}$. Appropriate cuts can be associated with the elements of this subfield to produce an isomorphism with $\mathcal{R}$.)

Where we speak of an isomorphic embedding, Euclid spoke of the "comparability" of two Archimedean systems of magnitudes. Obviously (in antiquity) it would have been conceptually difficult to discuss whether the positive axis contained "enough" or "all" points. Nevertheless, as Archimedes' method of exhaustion shows, the idea of completeness was present, if latent.

Why the step to the real numbers was not already taken in antiquity is a matter of speculation. One can give two reasons, one philosophical, the other mathematical, both of which influenced the later development of mathematics.

- *The Atomistic Tendency.* The idea of a smallest indivisible entity, an idea which originated in Aristotelean philosophy, reappears again and again in the history of the foundations of analysis. Even Galileo and Cavalieri thought it necessary to use *indivisibili* in the justification of their quadratures.

- *The Existence of Non–Archimedean Systems of Magnitudes.* Already in ancient times there where examples of such systems, of quantities that one would certainly consider measurable and comparable but which are not Archimedean. Their proportions $(a : b)$ cannot therefore be isomorphically embedded in $\mathcal{R}$.

Horn Angles

Consider an angle between two circular arcs which meet at a point P in the plane:

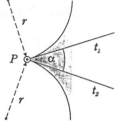

Fig. 1.1.

A horn angle is a pair $(\alpha, \frac{1}{r})$ where α is the usual angle between the tangents t_1 and t_2, and r is the radius of the two circles, as shown. The size of horn angles can be compared by "fitting them into one another", formally:

$$(\alpha, k) \leq (\beta, \ell) \equiv_{df} (\alpha < \beta) \vee (\alpha = \beta \wedge k \leq \ell)$$
$$(\alpha, k) + (\beta, \ell) =_{df} (\alpha + \beta, k + \ell).$$

Observe that $n \cdot (0, 1) < (\frac{\pi}{4}, 0)$ for all n.

The existence of systems of magnitudes that are apparently measurable but which cannot be measured by real numbers was considered paradoxical by the Greeks. There is definite evidence for this in remarks of Zeno of Elea, Protagoras, Democritius, and Eudoxus.

Much later, Vieta and Galileo in the sixteenth and Wallis in the seventeenth century concerned themselves with the problem. In modern measure theory it is disregarded: the range of values of a measure μ is invariably taken to be $\mathbb{R}$. Econometrics might provide an exception; perhaps "utilities" should be measured by non–archimedean magnitudes? Non–Archimedean ordered fields will be discussed in Section 4, even some which satisfy all the axioms of the elementary theory of the real numbers!

Now it is time to present a second axiomatization. This one relies upon the notion of a real–valued function and can thus be considered an axiomatization of elementary analysis.

Elementary Analysis

Intended Structure: $\mathcal{A} = \langle \mathbb{R}, \mathbb{R}^{\mathbb{R}}, +, \cdot, -, {}^{-1}, 0, 1, \leq \rangle$

Language and Logic: Two–sorted first–order predicate calculus with equality; individual variables: $x, y, z, \ldots$; function variables: $f, g, \ldots$; individual constants: $0, 1$; symbols for operators on individuals: $+, \cdot, -, {}^{-1}$; function application: $\cdots(\cdots)$; basic predicate: $\leq$.

Axioms:

(i)   Axioms for Ordered Fields.

(ii) Axiom of Extensionality:

$$f = g \equiv \forall x (f(x) = g(x)).$$

(iii) Axiom of Completeness:

$$\forall f [\exists y \forall x (f(x) \leq y) \supset \exists z (\forall x (f(x) \leq z) \wedge$$

$$\forall y (\forall x (f(x) \leq y) \supset z \leq y))].$$

(iv) Axioms of Comprehension:

$$\forall x \, \exists! y \, A(x, y) \supset \exists f \forall x A(x, f(x)).$$

In this theory the strength of the Completeness Axiom is not tied to defining properties of real numbers, but to the notion of function. This notion does not have a unique interpretation; its present–day intuitive meaning has been reached as the result of a long and difficult development. Under the pretext of geometrical purity, Descartes insisted with all his authority that only algebraic functions should be considered legitimate. In our axiomatization this corresponds roughly to the restriction that the Comprehension Axiom, (iv), applies only to formulas $A(x, y)$ in which no function variables occur. This restriction soon proved too strong; the sine function was one of the first shown to be transcendental. Thereafter the notion of function was extended (often amid heated discussion) to include antiderivatives, solutions to differential equations, power series expansions, etc. It can be shown that the existence of such functions follows from our axioms when $A(x, y)$ may be taken to be any formula in the language. (We do not wish to go into it here.) Ultimately historical development; i.e., simply the weight of the applications, the drawing of functions out of "nature", led to the modern notion of function — "an arbitrary function".

How definite is this notion? It can clearly be a subject of controversy. We can ask, for example: can one, should one add the Axiom of Choice to our theory?

(v)   *Axioms of Choice:*

$$\forall x \, \exists y \, A(x,y) \supset \exists f \, \forall x \, A(x, f(x)).$$

Are there other principles that one should or could add, principles which should enrich the theory with further interesting theorems. What about the Continuum Hypothesis etc.? We shall return to this question in Section 5.

Further Reading

Frege & Russel: Briefwechsel (Widerspruch in der Mengenlehre), appeared in: Jean van Heijenoort: From Frege to Gödel, a source book in mathematical logic 1879-1931, pp. 124-128, Cambridge, Mass., Harvard University Press, 1967

Bernays, P.: Betrachtungen über das Vollständigkeitsaxiom und verwandte Axiome, Mathematische Zeitschrift, vol. 63, pp. 219-299, (1955)

Euklid: Die Elemente, Ostwalds Klassiker der exakten Wissenschaften, Nr. 235, 1. Teil, 3. Buch, §16, pp. 57-59, und Nr. 236, 2. part, 5. book, pp. 17-36. Leipzig, Akad. Verlagsgesellschaft, 1932

Archimedes: The Works of Archimedes with the Method of Archimedes, edited by T.L. Heath, in particular the introduction in the appendix pp. 5-11, also pp. 12-51. New York. Dover

Kasner, E.: The Recent Theory of the Horn Angle, Scripta Mathematica, vol. 11, pp. 263-267, (1945)

Galilei: Galilei on Infinites and Infinitesimals, in: D.J. Struik: A Source Book in Mathematics, 1200-1800, pp. 198-207. Cambridge, Mass., Harvard University Press, 1969

§3   Elementary Theory of Real Numbers

The axiomatization presented in the previous section is an attempt to axiomatize the set of theorems true in the intended structure. How well does it do this? The best for which one can hope in an axiomatization is that it can serve as the basis for an effective decision procedure, i.e. a procedure which, given any sentence S of the language, decides in finitely many steps whether or not S is true in the intended structure. The most informative decision procedures are those which proceed by quantifier elimination. This method is also the oldest. It was applied in the twenties by Langford to the theory of dense orderings, by Presburger to the (additive) theory of the integers, and finally by Tarski to the theory that we are considering. The paper by Yu. Ershov, mentioned in the references, surveys numerous subsequent applications of the method.

Quantifier Elimination

Given: A first-order theory.

Assumption: For every formula A of the form $\exists x(A_1(x) \wedge \ldots \wedge A_n(x))$, where the $A_i(x)$ are negated or unnegated atomic formulas, there exists a quantifier-free formula B such that the equivalence $A \equiv B$ can be proved in the theory.

Elimination procedure: Given a sentence S

(1) Change the innermost quantifiers into existential quantifiers, if they are not already existential.

(2) Put the scope of each of these quantifiers into disjunctive normal form.

(3) Distribute the existential quantifiers over the disjuncts.

(4) Eliminate them by using the assumption.

(5) If the resulting expression is quantifier–free, evaluate it as true or false; otherwise repeat this procedure starting from (1).

The method of quantifier elimination reduces the logical decision problem to a mathematical one concerning criteria for the existence of solutions to specific problems. Such questions suggest themselves to mathematicians, so it is not surprising that there were problems in the classical theory of real algebra that can be formulated as questions about existence criteria. The following is a case in point.

Theorem (Sturm, 1829) *Let $p(x)$ be a given polynomial with integral coefficients. Construct a finite sequence of polynomials $p_1(x), p_2(x), \ldots, p_r(x)$ with $p_1(x) = p'(x)$, the derivative of p, and the others obtained by the Euclidean algorithm:*

$$
\begin{aligned}
p(x) &= q_1(x) \cdot p_1(x) - p_2(x), \\
p_1(x) &= q_2(x) \cdot p_2(x) - p_3(x), \\
&\vdots \\
p_{r-1}(x) &= q_r(x) \cdot p_r(x).
\end{aligned}
$$

For each a, define the Sturm chain to be

$$p(a), p_1(a), p_2(a), \ldots, p_r(a)$$

and let $\omega(a)$ be the number of changes of sign in this chain. Then, for $b < c, \omega(b) - \omega(c)$ is the number of distinct roots of $p(x)$ in the interval (b, c).

In the light of what is required for quantifier elimination, we can reformulate this theorem as follows:

Sturm's Theorem *For every polynomial $p(x, x_1, \ldots, x_n)$ with integral coefficients, there exists a quantifier–free formula $B(x, x_1, \ldots, x_n, a, b)$ such that*

$$a < b \supset \; . \; B(x_1, \ldots, x_n, a, b) \equiv \exists x (a < x < b \wedge p(x, x_1, \ldots, x_n) = 0)$$

follows from the axioms of the elementary theory of the real numbers.

The Generalized Sturm Theorem *For every quantifier-free formula $A(x, x_1, \ldots, x_n)$ there is a quantifier–free $B(x, x_1, \ldots, x_n, a, b)$ for which*

$$a < b \supset \; . B(x_1, \ldots, x_n, a, b) \equiv \exists x (a < x < b \wedge A(x, x_1, \ldots, x_n))$$

follows from the axioms of the elementary theory of the real numbers.

In algebra, Sturm's Theorem is usually proved by using analysis; in fact by appealing to the Weierstrass Theorem, according to which any continuous function that changes sign has a zero. The purely algebraic character of Sturm's

---

Elementary Theory of Real Closed Fields

Language and Logic: First–order Predicate Calculus with equality; individual variables: $x, y, z, \ldots$; individual constants: $0, 1$; function symbols: $+, \cdot, -, {}^{-1}$; basic predicate: $\leq$.

Axioms:

  (i)   Axiom for fields;

  (ii) Order axioms;

 (iii)′ $\forall x \exists y (x = y^2 \vee -x = y^2)$;

 (iii)″′ For each natural number n, the axiom

$$\forall x_0 \forall x_1 \ldots \forall x_{2n} \exists y (x_0 + x_1 \cdot y + x_2 \cdot y^2 + \ldots + x_{2n} \cdot y^{2n} + y^{2n+1} = 0)$$

Theorem and similar results suggested that it should be possible to prove them by purely algebraic means, without applying continuity arguments: to lay bare the fundamental algebraic facts underlying these results. This is the program that was completed by Artin and Schreier in their famous article. It happens that the consequences (iii)′ and (iii) of the Completeness Axiom (iii) suffice for the Artin–Schreier program.

Now we shall first of all show that the quantifier elimination follows from the Generalized Sturm theorem, and then we shall prove the theorem itself using elementary knowledge of the algebra of real numbers (such as the Weierstrass and Rolle theorems). That these techniques are generally available in the theory of real closed fields follows from the Artin-Schreier program (as expounded in § 79 of van der Waerden's book of Algebra).

1. Quantifier Elimination as a Consequence of the Generalized Sturm Theorem

Let $A(x, x_1, \ldots, x_n)$ be a quantifier–free formula, and suppose we wish to eliminate the existential quantifier $\exists x$ in $\exists x A(x, x_1, \ldots, x_n)$. Clearly, in every ordered field, we have

$$\begin{aligned}
\exists x A(x, x_1, \ldots, x_n) \equiv\ & A(-1, x_1, \ldots, x_n) \vee A(1, x_1, \ldots, x_n) \\
& \vee \exists x(-1 < x < 1 \wedge A(x, x_1, \ldots, x_n)) \\
& \vee \exists x(-1 < x < 0 \wedge A(x^{-1}, x_1, \ldots, x_n)) \\
& \vee \exists x(0 < x < 1 \wedge A(x^{-1}, x_1, \ldots, x_n))
\end{aligned}$$

The existential quantifiers can be eliminated from the formula on the right as a consequence of the theorem, which we shall now prove.

2. Proof of the Generalized Sturm Theorem

Our presentation of the proof follows a pattern due to P. Cohen; a similar one may be found in the book by Kreisel and Krivine. The essential notion in the proof is the degree of a quantifier–free formula. The atomic formulae of our language can clearly be interpreted as equations or inequalities between rational functions with integer coefficients. They are thus equivalent (modulo field theory) to equations $p = 0$ and inequalities $q > 0$ for polynomials p, q with integer coefficients. The degree of a quantifier–free formula $A(x)$ is defined to be $\max\{\text{degree}(p), \text{degree}(q)+1 \mid p = 0 \text{ or } q > 0 \text{ occurs as an atomic formula in } A\}$.

Lemma 1   *Let* $p_1, \ldots, p_k, q_1, \ldots, q_\ell$ *be polynomials in* $x, x_1, \ldots, x_m$ *with integer coefficients. Then*

$$p_1 = 0 \wedge p_2 = 0 \wedge \ldots \wedge p_k = 0 \wedge q_1 > 0 \wedge \ldots \wedge q_\ell > 0$$

is equivalent to a quantifier–free formula whose degree in x *is less than the degree in* x *of each polynomial* p_i.

Proof. Let h be the sum of the degrees of the p_i's and q_j's. We use induction on h:

$h = 0$:   Nothing to prove.

$h > 0$:   We argue by cases according to the value of k.

$k = 0$:   Nothing to prove.

$k = 1$:   The conjunction is of the form

$$p = 0 \wedge q_1 > 0 \wedge q_2 > 0 \wedge \ldots \wedge q_\ell > 0. \qquad (*)$$

The only interesting case arises when, for example, degree $(q_1) \geq$ degree(p). Let $p = ax^m + \ldots$, and $q = bx^n + \ldots$, where $a, b \neq 0$ and $m \leq n$. Set $Q = a^2 q_1 - abx^{n-m} \cdot p$. Then the conjunction $(*)$ is equivalent to

$$p = 0 \wedge Q > 0 \wedge q_2 > 0 \wedge \ldots \wedge q_\ell > 0.$$

To see this, suppose $p = 0 \wedge q_1 > 0$. Then $p = 0$ and $Q = a^2 q_1 > 0$ (since $a \neq 0$).Conversely, if $p = 0$ and $Q > 0$ then $a^2 q_1 > 0$, and so $q_1 > 0$. Moreover, degree$(Q) <$ degree(q_1). Thus we have reduced the sum of the degrees h and can therefore apply the induction assumption.

$k \geq 2$:   Take $p_1 = a_1 x^{m_1} + \ldots, p_2 = a_2 x^{m_2} + \ldots$, with $a_1, a_2 \neq 0$ and $m_1 \geq m_2$. Let $P = a_2 p_1 - a_1 x^{m_1 - m_2} p_2$. Then just as in the case $k = 1$, the given formula is equivalent to

$$P = 0 \wedge p_2 = 0 \wedge \ldots \wedge p_k = 0 \wedge q_1 > 0 \wedge \ldots \wedge q_\ell > 0.$$

Again the sum h has been decreased, and we can use the induction assumption.   □

Lemma 2 *Let $A(x, x_1, \ldots, x_n)$ be a quantifier–free formula of degree h in x. Choose a, b to be parameters distinct from $x, x_1, \ldots, x_n$, that do not occur in A. Then there exists a quantifier-free formula $B(x_1, \ldots, x_n, a, b)$ satisfying*

$$a < b \supset . B(x_1, \ldots, x_n, a, b) \equiv \exists x (a < x < b \wedge A(x, x_1, \ldots, x_n)).$$

Moreover, the degree of B in a and b is bounded by $h + 1$, and each atomic formula occurring in B contains at most one of the parameters a, b.

Proof. By induction on the degree h of A in x.

$h = 0$:   In this case the variable x does not occur in $A(x, x_1, \ldots, x_n)$, and so we can take B to be A itself.

$h > 0$:   Without loss of generality we can suppose that A is of the form

$$p_1 = 0 \wedge p_2 = 0 \wedge \ldots \wedge p_k = 0 \wedge q_1 > 0 \wedge \ldots \wedge q_\ell > 0,$$

since we may assume that A has been converted into disjunctive normal form and that the existential quantifier has been distributed over the disjuncts.

We distinguish between cases for k.

$k = 0$:　A is of the form

$$q_1 > 0 \wedge \ldots \wedge q_\ell > 0.$$

The statement

$$\exists x(a < x < b \wedge q_1 > 0 \wedge \ldots \wedge q_\ell > 0),$$

then requires that there is some point at which (and therefore some subinterval $(\alpha, \beta) \subseteq (a, b)$ on which) all the polynomials $q_1, \ldots, q_\ell$ are strictly positive. This can happen in a number of ways. Either

I.　$G_0(a, b) \equiv \forall x(a < x < b \supset . q_1 > 0 \wedge \ldots \wedge q_\ell > 0),$

or there is an $i, 1 \leq i \leq \ell$, such that

II.　$G_i(a, b) \equiv \exists v(a < v < b \wedge q_1(v) = 0 \wedge G_0(a, v))$
$\vee \exists v(a < v < b \wedge q_i(v) = 0 \wedge G_0(v, b)),$

or there exist $i, j, 1 \leq i \leq \ell$ and $1 \leq j \leq \ell$, such that

III.　$H_{ij}(a, b) \equiv \exists u \exists v(a < u < v < b \wedge q_i(u) = 0$
$\wedge q_j(v) = 0 \wedge G_0(u, v)).$

These three situations in the elimination of quantifiers from $\exists x(a < x < b \wedge q_1 > 0 \wedge \ldots \wedge q_\ell > 0)$ can be depicted graphically.

I.　$G_0(a, b) : \forall x(a < x < b \supset . q_1 > 0 \wedge \ldots \wedge q_\ell > 0),$

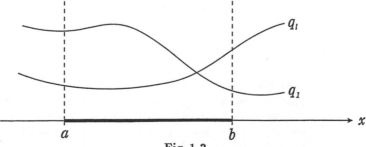

Fig. 1.2.

II. $G_i(a,b): \quad \exists v(a < v < b \wedge q_1(v)) = 0 \wedge G_0(a,v))$
$$\vee \exists v(a < v < b \wedge q_i(v)) = 0 \wedge G_0(v,b)),$$

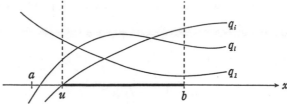

Fig. 1.3.

III. $H_{ij}(a,b): \quad \exists u \exists v(a < u < v < b \wedge q_i(u) = 0$
$$\wedge q_j(v) = 0 \wedge G_0(u,v)).$$

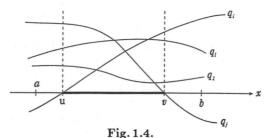

Fig. 1.4.

It therefore follows that

$$\exists x(a < x < b \wedge q_1 > 0 \wedge \ldots \wedge q_\ell > 0) \equiv$$
$$G_0(a,b) \vee G_1(a,b) \vee \ldots \vee G_\ell(a,b) \vee H_{11}(a,b) \vee \ldots$$
$$\vee H_{1\ell}(a,b) \vee H_{21}(a,b) \vee \ldots \vee H_{\ell\ell}(a,b).$$

So there are $\ell^2 + \ell + 1$ disjuncts to be reduced.

I.   *Reduction of* $G_0(a,b)$:

$$\begin{aligned} G_0(a,b) \quad &\equiv \quad \forall x(a < x < b \supset .q_1 > 0 \wedge \ldots \wedge q_\ell > 0) \\ &\equiv \quad \forall x(a < x < b \supset q_1 > 0) \\ &\qquad \wedge \forall x(a < x < b \supset q_2 > 0) \\ &\qquad \vdots \\ &\qquad \wedge \forall x(a < x < b \supset q_\ell > 0) \end{aligned}$$

Note that, since $h = \text{degree}(A)$, the degree of each of the polynomials $q_1, \ldots, q_\ell$ is strictly less than h. Moreover, a polynomial q is strictly positive on the open interval (a,b) precisely in the case q has no zero in the interval and either q itself or the first non–zero derivative of q at a is positive.

Formally,

$$\forall x(a < x < b \supset q_i > 0) \equiv \neg \exists x(a < x < b \wedge q_1 = 0)$$
$$\wedge (q_i(a) > 0$$
$$\vee (q_i(a) = 0 \wedge q_i'(a) > 0)$$
$$\vdots$$
$$\vee (q_i(a) = 0$$
$$\wedge q_i'(a) = 0 \wedge q_i'(a) =$$
$$0 \wedge \ldots \wedge q_i^{(h-2)}(a) = 0$$
$$\wedge q_i^{(h-1)}(a) = 0)).$$

For each $i, 1 \leq i \leq \ell$, the degree in x of each of these formulae is strictly less than h. By the induction assumption each of these formulae can therefore be reduced. Hence we can also reduce the conjunction of ℓ such formulae, namely $G_0(a, b)$. As a result of the reduction we obtain a formula of the type

$$\bigwedge_{i=1}^{\ell} (\neg B_i(x_1, \ldots, x_n, a, b) \wedge K_i(a)).$$

This has the form

$$K(a) \wedge L(b)$$

and, by the induction assumption, its degree in a and b is bounded by h.

II.   *Reduction of $G_i(a, b), 1 \leq i \leq \ell$:*

$$G_i(a, b) \equiv \exists v(a < v < b \wedge q_i(v) = 0 \wedge G_0(a, v))$$
$$\vee \exists v(a < v < b \wedge q_i(v) = 0 \wedge G_0(v, b)).$$

By I. we have

$$G_0(a, v) \equiv K(a) \wedge L(v)$$
$$G_0(v, b) \equiv K(v) \wedge L(b)$$

from which it follows that

$$G_i(a, b) \equiv K(a) \wedge \exists v(a < v < b \wedge q_i(v) = 0 \wedge L(v))$$
$$\vee \exists v(a < v < b \wedge q_i(v) = 0 \wedge K(v) \wedge L(b)).$$

By what has been said above, h is an upper bound on the degree of $G_i(a, b)$ in v. Lemma 1 can be applied to decrease the degree of each disjunct to at most $h - 1$. Hence we can apply the induction assumption to $G_i(a, b)$, which is therefore reducible.

III.   *Reduction of* $H_{ij}(a,b), 1 < i, j < \ell$:

$$H_{ij}(a,b) :\equiv \exists u \exists v (a < u < b \wedge q_i(u) = 0$$
$$\wedge q_j(v) = 0 \wedge G_0(u,v)).$$

By I., $G_0 \equiv K(u) \wedge L(v)$. Therefore

$$H_{ij}(a,b) \equiv \exists u (a < u < b \wedge q_i(u) = 0 \wedge K(u)$$
$$\wedge \exists v (u < v < b \wedge q_j(v) = 0 \wedge L(v))).$$

The degree in v of the inner existentially quantified formula is bounded by h (as consequence of I.); moreover, degree$(q_j) \leq h - 1$. Lemma 1 can be used to decrease the degree in v of this inner formula to the degree of q_j thus enabling us to apply the induction assumption. As a result we obtain

$$H_{ij}(a,b) \equiv \exists u (a < u < b \wedge q_i(u) = 0 \wedge K(u) \wedge M(u,b)).$$

The degree in u of this existentially quantified formula is bounded by h; Lemma 1 can be used to reduce this to the degree of q_i (which is at most $h - 1$) so that we can once again apply the induction assumption. $H_{ij}(a,b)$ can therefore be reduced.

$k = 1$:   In this case A is of the form

$$p = 0 \wedge q_1 > 0 \wedge \ldots \wedge q_\ell > 0.$$

Without loss of generality we can assume that

$$\text{degree}(p) = h, \quad \text{and degree}(q_i) < h, \text{ for } i = 1, \ldots, \ell.$$

(If this were not the case, we could use Lemma 1 to decrease the degree of p (less than h) and apply the induction assumption.) We have

$$\exists x (a < x < b \wedge p = 0 \wedge q_1 > 0 \wedge \ldots \wedge q_\ell > 0) \equiv$$
$$\exists x (a < x < b \wedge p = 0 \wedge q_0 > 0 \wedge \ldots \wedge q_\ell > 0) \vee$$
$$\exists x (a < x < b \wedge p = 0 \wedge q_0 = 0 \wedge \ldots \wedge q_\ell > 0) \vee$$
$$\exists x (a < x < b \wedge p = 0 \wedge -q_0 > 0 \wedge \ldots \quad \wedge q_\ell > 0),$$

where q_0 denotes the derivative of the polynomial p. Let us call the three alternatives $A_1, A_2,$ and A_3.

I.   *Reduction of* A_1. Once again we consider the formulae

$$G_0(a,b) \equiv \forall x(a < x < b \supset .q_0 > 0 \wedge \ldots \wedge q_\ell > 0)$$
$$G_i'(a,b) \equiv \exists u(a < u < b \wedge q_i(u) = 0 \wedge p(u) > 0 \wedge G_0(a,u))$$
$$G_i''(a,b) \equiv \exists v(a < v < b \wedge q_i(v) = 0 \wedge -p(v) > 0 \wedge G_0(v,b))$$
$$H_{ij}(a,b) \equiv \exists u \exists v(a < u < \ v < b \wedge q_i(u) = 0 \wedge q_j(v) = 0$$
$$\wedge - p(u) > 0 \wedge p(v) > 0 \wedge G_0(u,v))$$

By the Weierstrass Theorem and the theorem that a function with positive derivative is monotonic increasing, it follows that

$$A_1 \equiv (-p(a) > 0 \wedge p(b) > 0 \wedge G_0(a,b))$$
$$\vee(-p(a) > 0 \wedge G_0'(a,b))$$
$$\vdots$$
$$\vee(-p(a) > 0 \wedge G_\ell'(a,b))$$
$$\vee(p(b) > 0 \wedge G_0''(a,b))$$
$$\vdots$$
$$\vee(p(b) > 0 \wedge G_\ell''(a,b))$$
$$\vee H_{00}(a,b) \vee \ldots \vee H_{\ell\ell}(a,b).$$

As in the case $k = 0$, one shows that

$$G_0(a,b) \equiv K(a) \wedge L(b),$$

where the degrees of $K(a)$ and $L(b)$ in a and b are bounded by h. Again it follows from this that G_i', G_i'', and H_{ij} can be reduced. Hence A_1 can be reduced to a formula whose degree in a and b is bounded by $h+1$ (since $p(a) > 0$ is of degree $h+1$ in a).

II.   *Reduction of A_2.* We apply Lemma 1 to obtain a formula of degree $h - 1(= \text{degree}(q_0))$; by the induction assumption, this formula is reducible.

III. *Reduction of A_3.* This is similar to the reduction of A_1.

$k = 2:$ The Euclidean algorithm can be used to reduce this to the case $k = 1$. Indeed, $p_1 = 0$ and $p_2 = 0$ have a common solution if and only if their greatest common divisor (whose degree is certainly not greater than that of p_1 or p_2) has a zero. $\square$

What we have now proved, that quantifier elimination can be applied in theory of real closed fields, can be regarded in a number of different ways. The most obvious is to say that with this theorem high-school algebra reaches its goal: in principle, anyone who knows the decision procedure need not learn any more about the algebra of the real numbers. We shall return to this later. At a more abstract level, one can view the theorem both from a mathematical and from a formal logical point of view. First let us remark that, by iterating the quantifier elimination procedure, we can clearly eliminate any finite sequence of existential quantifiers and not just one; in particular the formula

$$\exists x_1 \exists x_2 \ldots \exists x_n (p_1 = 0 \wedge \ldots \wedge p_k = 0 \wedge q_1 > 0 \wedge \ldots \wedge q_\ell > 0)$$

can be transformed into an equivalent quantifier–free disjunction of systems of polynomial equations and inequalities. To do so we would have to repeat the quantifier elimination procedure described above n times. Seidenberg found an alternative procedure for quantifier elimination. It uses methods from algebraic geometry and enables one to eliminate arbitrarily many existential quantifiers directly and at once. Moreover, in view of the equivalence

$$p > 0 \equiv \exists v (p \cdot v^2 = 1),$$

inequalities do not have to be considered. Furthermore, since

$$p_1 = 0 \wedge \ldots \wedge p_k = 0 . \equiv p_1^2 + p_2^2 + \ldots + p_k^2 = 0,$$

it remains only to show that, for each polynomial p in n variables it can be decided, whether or not p has a zero in $\mathbb{R}^n$, i.e. whether or not

$$\exists x_1 \exists x_2 \ldots \exists x_n (p = 0).$$

Tarksi's Theorem (Mathematical Form) *Given any system S of equations and inequalities of rational functions in $x_1, \ldots, x_n$ with parameters $a_1, \ldots, a_m$, there is an effective procedure for generating finitely many systems $T_1, \ldots, T_k$ of polynomial equations and inequalities such that S has a solution in $x_1, \ldots, x_n$ if and only if the parameters $a_1, \ldots, a_m$ satisfy at least one of the systems T_i.*

This theorem is the completion of real algebra in the sense that it suggests a secure method of solving any given problem, and doing so successfully. It has to be realized, however, that the very generality of the solution procedure prejudices its efficiency; for any particular class of problems it is definitely preferable to aim for a more specific algorithm. The general theorem has beautiful theoretical applications. To mention just one example, A. Friedman's book describes an application to partial differential equations due to P.C. Rosenbloom.

Is the decision procedure itself useful? So–called elementary mathematics is a source of many fascinating problems, some still unsolved, to which the decision procedure can in principle be applied. We shall mention two of these, one due to Poncelet (the founder of projective geometry), the other to Euler.

Poncelet's Problem

Given two conic sections in the plane, is there a polygonal path which is circumscribed by one and inscribed by the other?

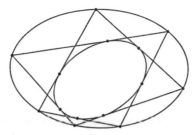

Fig. 1.5.

Poncelet himself solved this problem, and later both Jacobi and Cayley concerned themselves with it. (For a more recent work on the problem, see the article by P. Griffiths and J. Harris in L'Enseignement Mathématiques 26 (1978).)

Euler's Problem

How many points at a distance at least r from one another can there be on the surface of a sphere of radius r?

Twelve or thirteen?

This problem was not solved until the present century (by Schütte and van der Waerden), and then with analytic techniques. We shall use this example to illustrate a rather typical phenomenon. Euler's conjecture can clearly be expressed by a formula of very low degree. As we have seen, the proof by quantifier elimination will never introduce formulae of higher degree. By contrast, the shorter analytic proof uses estimates obtained by means of trigonometric functions. Any such proof can be carried out in the elmementary theory by replacing the trigonometric functions by the initial partial sums of their power series expansions. But the estimates can only work if these intial sums are of higher degree than the polynomials appearing in the statement of the problem!

What about the realization of a decision procedure on a computer? If it is to serve any practical purpose, such a project would be extremely large and demanding. The only system with any real promise at present is that of G. Collins, in Wisconsin. As Tarski already remarked, the central question concerns the complexity of the decision algorithm with respect to the size of the given formula. The following result, which we state without proof, is relevant to this subject.

Fischer–Rabin Theorem   *There is a constant $d > 0$ such that for any decision algorithm for sentences of real algebra that do not involve multiplication or inverses there exists a number n_0 with the following property: for any natural number $n \geq n_0$ there is a sentence of length n which the algorithm requires at least $2^{d \cdot n}$ computational steps to decide.*

(The computational complexity of Collins' procedure is $(2n)2^{2r+8}m^{2r+6}d^3e$, where r is the number of variables, m the number of polynomials, n the highest degree of the polynomials, d the length of the longest integer coefficient, and e the number of atomic formulae occuring in the given sentence. For a sentence of length n, this corresponds roughly to an execution time of $2^{2^{k \cdot n}}$, where $k \leq 8$.)

Now let us examine Tarski's Theorem from the point of view of formal logic. As we have seen, an analysis of the proof reveals that the only properties of the field of real numbers that are used are those which follow from the axioms for real closed fields.

Tarksi's Theorem (Logical Form)   *A theorem of elementary algebra holds in the field of real numbers if and only if it holds in every real closed field. Validity, i.e. derivability from the axioms of real closed fields, is decidable.*

This theorem settles a whole list of questions about field theory in an almost trivial way. Take for example Brouwer's Fixed Point Theorem: let M be a closed, bounded convex subset of $\mathbb{R}^n$ and let f be a continuous mapping from M to itself; then f has a fixed point in M.

We specialize this theorem so as to convert it to a theorem of real algebra: take M to be the set of points $\vec{x}$ that satisfy a given formula $A(\vec{x})$. The theorem can be transformed as follows:

M is bounded:
$$\exists z(z \geq 0 \land \forall \vec{x}(A(\vec{x}) \supset |\vec{x}| \leq z)).$$

M is closed:
$$\forall \vec{x}(\neg A(\vec{x}) \supset \exists w(w > 0 \land \forall \vec{z}(|\vec{x} - \vec{z}| < w \supset \neg A(\vec{z})))).$$

M is convex:
$$\forall \vec{x} \forall \vec{y}(A(\vec{x}) \land A(\vec{y}) . \supset \forall u \forall v(0 \leq u \leq 1 \land 0 \leq v \leq 1 \land u + v = 1 . \supset A(u \cdot \vec{x} + v \cdot \vec{y}))).$$

The rational function $f = \frac{p}{q}$ is continuous in M:
$$\forall \vec{x}(A(\vec{x}) \supset q(\vec{x}) \neq 0).$$

The function f has a fixed point in M:

$$\exists \vec{x}(f(\vec{x}) = \vec{x} \wedge A(\vec{x})).$$

Completely translated into this algebraic form, Brouwer's Theorem therefore holds in all real closed fields (in spite of the fact that the topological methods used in its proof are unavailable here).

The Brouwer–Tarski Theorem *Let M be an elementarily definable, closed, bounded, convex set in F^n, for some real closed field F. Then every continuous rational function mapping M to itself has a fixed point in M. (This can be generalized still further.)*

A further consequence of Tarski's Theorem is the Definability Theorem, which provides an answer to the question raised in Section 2 as to which sets are definable by properties $A(x)$ in the elementary language.

Definability Theorem *The sets definable in the elementary language of $\mathcal{R}$ are precisely those subsets $M \subseteq \mathbb{R}$ that consist of finite unions of open (half-open, closed) intervals with algebraic endpoints.*

An immediate corollary of this theorem, itself a consequence of quantifier elimination, is the fact that there is no elementary formula $N(x)$ that can be used to define the set $\mathbb{N}$ of natural numbers as a subset of $\mathbb{R}$.

Finally we mention A. Robinson's and Kreisel's application to the solution (Artin, 1927) of Hilbert's 17th Problem.

Hilbert's Seventeenth Problem *Let f be a rational function in n variables over $\mathbb{R}$, and assume $f > 0$ on $\mathbb{R}^n$. Do there exist rational functions $g_i, i = 1,\ldots,m$ such that $f = \sum_{i=1}^{m} g_i^2$?*

Artin's answer is positive. Robinson gave a more direct proof using methods of formal logic, and Kreisel shows that the number of m and the maximum degree of any polynomial occurring in one of the g_i can be computed given the maximum degree of the numerator and denominator of f. The computation uses the decision procedure; however, the bounds obtained this way for the general case are not the best possible.

Further Reading

Tarski, A.: A Decision Method of Elementary Algebra and Geometry, Berkeley, University of California Press, 1951

Seidenberg, A.: A New Decision Method for Elementary Algebra, Annals of Mathematics, vol. 60, pp. 365-374, (1954)

Ershov, Yu. et. al.: Elementary Theories, Russian Mathematical Surveys, vol. 20 (4), pp. 35-105, in particular pp. 94-99, (1965)

Fischer, M.J. & Rabin, M.O.: Super–Exponential Complexity of Presburger Arith-
    metic, SIAM-AMS Proceedings, vol. 7 pp. 27-41, (1974)
Collins, G.E.: Quantifier Elimination for Real Closed Fields by Cylindrical Algebra
    Decomposition, Springer Lecture Notes in Computer Science 33, pp. 134-183, (1975)
Hilbert, D.: Mathematische Probleme, 17. Problem: Darstellung definiter Formen
    durch Quadrate, Ostwalds Klassiker der exakten Wissenschaften, vol. 252, pp. 63-
    64. Leipzip, Akad. Verlagsgesellschaft, 1933
Artin, E. & Schreier, O.: Algebraische Konstruktion reeller Körper, Abhandlungen aus
    dem mathematischen Seminar der Universität Hamburg, vol. 5, pp. 85-99, (1927)
Artin, E.: Ueber die Zerlegung definiter Funktionen in Quadrate, Abhandlungen aus
    dem mathematischen Seminar der Universität Hamburg, vol. 5 pp. 100-115, (1927)
Friedmann, A.: Generalized Functions and Partial Differential Equations, in particular
    pp. 218-225. Englewood Cliffs, Prentice–Hall, 1963
Heintz, J., Roy, M.-F., Solernó, P.: On the complexity of semialgebraic sets, Proc.
    IFIP (San Francisco 1989), North Holland, pp. 293-298, (1989)

§ 4 Non–Standard Analysis

For the first hundred and fifty years of its existence, differential and integral
calculus were known as the analysis of the infinitely small. Euler's influential
textbook, for example, is entitled "Introductio in Analysin Infinitorum" (Lau-
sanne, 1748). The infinitely small magnitudes which we encountered as "atoms
of straight lines" in Cavalieri's integration continue to play an important role.
The precursors of Newton and Leibniz found a new use for vanishingly small
quantities in problems of determining tangents and in finding maxima and minima. Then they were introduced systematically by Leibniz in the form of differentials. Throughout he intended that differentials should be understood as
legitimate elements of the range and domain of functions, but neither he nor
his successors could provide them with a solid mathematical foundation. Long
into the Enlightenment, analysis (through its stormy development) lived with
this somewhat makeshift and precarious notion among its basic concepts. In
his witty and well–informed polemics Bishop Berkeley could accuse the intel-
lectually arrogant scientists themselves of harboring dubious assumptions. Very
much to the point, he referred to differentials as "ghosts of departed quantities".

The nineteenth century saw a complete and methodical rejection of the infinitely small. For justification analysis turned to the $\epsilon - \delta$–methods of Cauchy
and Weierstrass that are still taught today. This approach lacks much of the
intuitive – albeit insecure – appeal of infinitesimals. It even led to the general belief that it was mathematically impossible to rescue the notion of the
infinitesimal and to return to Leibniz's paradise.

The discovery of non–Archimedean ordered fields in the twentieth century
paved the way for the rehabilitation of infinitesimals. But it was only in 1960
that A. Robinson discovered a clean, elementary way to reintroduce infinitesimals as bona–fide elements in the underlying domain of analysis.

Are the axioms of the elementary theory of real numbers categorical? That is, do other models exist besides the intended structure $\mathcal{R}$? That they do follows from general theorems of first–order–predicate calculus, in particular from the famous "Löwenheim–Skolem–Malcev–Tarski–Theorem". We shall give a direct proof of this fact below. How has this happened? Clearly while the restriction of the notion of continuity to elementarily definable sets has the desirable effect of making the theory complete and decidable, it also, so to say, loses sight of the intended structure.

The point of this section is to make a "virtue" out of "necessity" by actually exploiting the existence of unintended models for the real numbers. Such models cannot of course distinguish themselves from $\mathcal{R}$ in any properties that can be formulated in an elementary way, but they can certainly differ in other respects, for example topologically. The methods of so–called "non–standard analysis" gain their power from the subtle exploitation of this fact.

We start with the construction of models for the preceding theory of $\mathcal{R}$. As in normal in modern mathematics, "construction" here includes something thoroughly non–constructive. In particular, we shall freely make use of the following non–constructive existence theorem.

Zorn's Lemma *Let P be a partially ordered set in which every nonempty totally ordered subset has an upper bound. Then P has a maximal element.*

It is well known that Zorn's Lemma is equivalent to the Axiom of Choice, which we shall discuss further in Section 5. A useful technique in algebra is the construction of the direct power of a given algebraic structure, such as a group. The direct power of a group is itself always a group. However, the construction of the direct power of a structure may produce a structure with drastically different properties, as the following example shows.

Example of a Direct Power

Given structure $\mathcal{A}$:   A set of two elements a and b ordered by $a \leq b$ with $a \neq b$.

Direct power $\mathcal{A}^2$:   The set of pairs (x, y), where x, y are equal to a or b, with the ordering given by
$(x, y) \leq (u, v) \equiv . \; x \leq u \wedge y \leq v$.

$\mathcal{A}$ is totally ordered, but $\mathcal{A}^2$ is not.

$$(b, b)$$
$$\leq \qquad \geq$$
$$(a, b) \qquad\qquad (b, a)$$
$$\geq \qquad \leq$$
$$(a, a)$$

The direct power construction cannot in general be used to produce new structures with the same elementary properties as those from which they are constructed. This purpose is served however by a refinement of the direct power construction first used by Th. Skolem. The first general description of this construction is due to J. Los. To present it we need the notion of filter (the same filters that appear in topology).

Filters and Ultrafilters

Let $I \neq \emptyset$, and let D be a nonempty family of subsets of I. D is said to be a filter on I if

(a) $\emptyset \notin D$;

(b) $E \in D \wedge F \in D . \supset E \cap F \in D$;

(c) $E \in D \wedge E \subseteq F \subseteq I . \supset F \in D$.

D is an ultrafilter on I, if in addition,

(d) $E \subseteq I \supset . E \in D \vee (I - E) \in D$.

An example of a filter is the family of those sets of natural numbers whose complements are finite. This is known as the Fréchet filter. It is obvious that this family of subsets of $\mathbb{N}$ is in fact a filter. It is not an ultrafilter however: it contains neither the set of even numbers nor the set of odd numbers. Nevertheless, the following general theorem shows that the Fréchet filter can be extended, non-constructively, to an ultrafilter.

Tarksi's Theorem *Every filter D_0 over I can be extended to an ultrafilter D over I.*

Proof.   Let F be the set of all filters D' over I that extend D_0, i.e. for which $D' \supseteq D_0$. Then F is not empty and is partially ordered by inclusion $\subseteq$. Every totally ordered set of filters in F has an upper bound, namely its union. Hence Zorn's Lemma can be applied, and F therefore has a maximal element, say, D. We claim that the filter D is in fact an ultrafilter. Suppose otherwise: Let $E \notin D$ and $(I - E) \notin D$. Construct the set D' of all $X \subseteq I$ for which $E \cup X \in D$. We can show that D' is a filter: $\emptyset \notin D'$, since $E \cup \emptyset = E \notin D$; if $E \cup X \in D$ and $E \cup Y \in D$ then $E \cup (X \cap Y) = (E \cup X) \cap (E \cup Y) \in D$; if $E \cup X \in D$ and $Y \supseteq X$, then $E \cup X \subseteq E \cup Y \in D$. By construction, $D' \supseteq D$, since $U \in D$ implies $U \subseteq E \cup U \in D$ and therefore $U \in D'$. Since D is maximal, $D = D'$. But this is impossible because $(I - E) \notin D$ while $(I - E) \in D'$ since $E \cup (I - E) = I \in D$.   □

Let D be an ultrafilter that extends the Fréchet filter over $\mathbb{N}$; we now come to the promised construction of ultrapowers. Intuitively, the idea is that an ultrapower of $\mathcal{R}$ should consist of sequences of reals, with two sequences being identified if they coincide "almost everywhere". The precise meaning of "almost everywhere" is taken over from the filter D. The sequences $\{a_i\}_{i\in\mathbb{N}}$ and $\{b_i\}_{i\in\mathbb{N}}$ should be considered equal if $\{i \in \mathbb{N} \mid a_i = b_i\} \in D$.

The meanings of "$\leq$", "$+$", etc. will be extended appropriately to apply to the equivalence classes thus defined. The details appear in the following box.

Ultrapower of $\mathcal{R}$

$\mathcal{R} = \langle \mathbb{R}, \leq, +, \cdot, -, ^{-1}, 0, 1 \rangle$.

$\mathbb{N} = \{0, 1, 2, \ldots\}$, D is an ultrafilter over $\mathbb{N}$ that contains all cofinite sets.

$\mathcal{R}_D^\mathbb{N} = \langle \mathbb{R}^\mathbb{N}/D, \leq_D, +_D, \cdot_D, -_D, {}_D^{-1}, 0_D, 1_D \rangle$, is defined by

$$\{a_i\}_{i\in\mathbb{N}}/D = \{\{b_i\}_{i\in\mathbb{N}} \mid \{i \in \mathbb{N} | a_i = b_i\} \in D\};$$
$$\mathbb{R}^\mathbb{N}/D = \{\{a_i\}_{i\in\mathbb{N}}/D \mid a_i \in \mathbb{R}, i \in \mathbb{N}\};$$
$$\{a_i\}_{i\in\mathbb{N}}/D \leq_D \{b_i\}_{i\in\mathbb{N}}/D \equiv \{i \in \mathbb{N} | a_i \leq b_i\} \in D;$$
$$\{a_i\}_{i\in\mathbb{N}}/D +_D \{b_i\}_{i\in\mathbb{N}}/D = \{a_i + b_i\}_{i\in\mathbb{N}}/D, \text{ etc.;}$$
$$\vdots$$
$$0_D = \{0,0,0,\ldots\}/D; \quad 1_D = \{1,1,1,\ldots\}/D.$$

We have to show that this definition of ultrapower is legitimate, and in particular that the operations $+_D$, etc, are well–defined on the equivalence classes. First let us convince ourselves that the identification of sequences is in fact an equivalence relation: $\{a_i\}_{i\in\mathbb{N}}/D = \{b_i\}_{i\in\mathbb{N}}/D$ if and only if $\{i \in \mathbb{N} \mid a_i = b_i\} \in D$. The relation is trivially reflexive and symmetric; transitivity follows from the fact that

$$\{i \in \mathbb{N} \mid a_i = b_i\} \cap \{i \in \mathbb{N} \mid b_i = c_i\} \subseteq \{i \in \mathbb{N} \mid a_i = c_i\}.$$

The ordering $\leq_D$ is well–defined because $\{a_i\}_{i\in\mathbb{N}}/D = \{a_i'\}_{i\in\mathbb{N}}/D$, $\{b_i\}_{i\in\mathbb{N}}/D = \{b_i'\}_{i\in\mathbb{N}}/D$, and $\{a_i\}_{i\in\mathbb{N}}/D \leq \{b_i\}_{i\in\mathbb{N}}/D$ together imply that $\{a_i'\}_{i\in\mathbb{N}}/D \leq_D \{b_i'\}_{i\in\mathbb{N}}/D$. To see this, let $E = \{i \in \mathbb{N} \mid a_i = a_i'\}$, $F = \{i \in \mathbb{N} \mid b_i = b_i'\}$, and $G = \{i \in \mathbb{N} \mid a_i \leq b_i\}$. By assumption each of these sets is in D. It follows that $\{i \in \mathbb{N} \mid a_i' \leq b_i'\} \supseteq E \cap F \cap G$ is in D. The operations $+_D, \cdot_D$, etc can similarly be shown to be well–defined. $\square$

The Skolem–Łos Theorem *Let* $A(x_1, \ldots, x_n)$ *be a formula in the elementary language of* $\mathbb{R}$ *with free variables* $x_1, \ldots, x_n$. *Replace these variables with any* n *elements* $\{a_i^j\}_{i\in\mathbb{N}}/D$, $j = 1, \ldots, n$. *Then the formula* $A(\{a_i^1\}_{i\in\mathbb{N}}/D, \ldots, \{a_i^n\}_{i\in\mathbb{N}}/D)$ *is true in* $\mathcal{R}_D^\mathbb{N}$ *if and only if*

$$\{i \in \mathbb{N} \mid A(a_i^1, a_i^2, \ldots, a_i^n) \text{ is true in } \mathbb{R}\} \in D.$$

In particular, the same elementary sentences are true in $\mathcal{R}$ and in $\mathcal{R}_D^{\mathbb{N}}$.

Proof. As always in such cases, the proof is by induction on the logical structure of $A(x_1, \ldots, x_n)$.

We shall use the abbreviations x for $x_1, \ldots, x_n$, a_i for $a_i^1, \ldots, a_i^n$, and a for $\{a_i\}_{i \in \mathbb{N}}/D$.

1. If $A(x)$ is an atomic formula then the statement of the theorem follows directly from the definition of the ultrapower $\mathcal{R}_D^{\mathbb{N}}$.

2. Suppose $A(x)$ is of the form $B(x) \wedge C(x)$. If $A(a)$ holds in $\mathcal{R}_D^{\mathbb{N}}$ then so do $B(a)$ and $C(a)$; so, by the induction assumption, both $E_1 = \{i \in \mathbb{N} \mid B(a_i) \text{ is true in } \mathcal{R}\}$ and $E_2 = \{i \in \mathbb{N} \mid C(a_i) \text{ is true in } \mathcal{R}\}$ are also in D. Then $E_1 \cap E_2 = E = \{i \in \mathbb{N} \mid B(a_i) \cap C(a_i) \text{ is true in } \mathcal{R}\}$ is also an element of D. Conversely, suppose $E \in D$. Then, since $E \subseteq E_1 \subseteq \mathbb{N}$ and $E \subseteq E_2 \subseteq \mathbb{N}$, both $B(a)$ and $C(a)$ are true in $\mathcal{R}_D^{\mathbb{N}}$, and therefore so is their conjunction.

3. Suppose $A(x)$ is of the form $\neg B(x)$. Then $A(a)$ holds in $\mathcal{R}_D^{\mathbb{N}}$ just in case $B(a)$ does not. By the induction assumption, this occurs just when $E = \{i \in \mathbb{N} \mid B(a_i) \text{ is true in } \mathcal{R}\}$ is not in D; since D is an ultrafilter this is the case if and only if the complement $(\mathbb{N} - E)$ does belong to D. But this complement can also be represented as $(\mathbb{N} - E) = \{i \in \mathbb{N} \mid \neg B(a_i)$ is true in $\mathcal{R}\}$.

4. Finally, suppose $A(x)$ is of the form $\exists y B(y, x)$. Assume $A(a)$ is true in $\mathcal{R}_D^{\mathbb{N}}$. Then there exists some element $b = \{b_i\}_{i \in \mathbb{N}}/D$ of $\mathbb{R}_D^{\mathbb{N}}$ such that $B(b, a)$ holds in $\mathcal{R}_D^{\mathbb{N}}$. The induction assumption now implies that $\{i \in \mathbb{N} \mid B(b_i, a_i) \text{ is true in } \mathcal{R}\} \in D$. But $\{i \in \mathbb{N} \mid B(b_i, a_i) \text{ is true in } \mathcal{R}\} \subseteq \{i \in \mathbb{N} \mid \exists y B(y, a_i) \text{ is true in } \mathcal{R}\}$. The latter set is therefore also in D.

   Conversely, assume that $\{i \in \mathbb{N} \mid \exists y B(y, a_i) \text{ is true in } \mathcal{R}\}$ belongs to D. We construct a sequence $\{b_i\}_{i \in \mathbb{N}}$ in $\mathbb{R}$ by choosing for each a_i an element b_i such that $B(b_i, a_i)$ holds in $\mathcal{R}$ if such an element exists, and choosing b_i arbitrarily if there is no such element. For each such sequence it is clearly the case that $\{i \in \mathbb{N} \mid B(b_i, a_i) \text{ is true in } \mathcal{R}\} \supseteq \{i \in \mathbb{N} \mid \exists y B(y, a_i) \text{ is true in } \mathcal{R}\}$, and so the larger set also belongs to D. It follows that $B(\{b_i\}_{i \in \mathbb{N}}/D, a)$, and therefore also $\exists y B(y, a)$ holds in $\mathcal{R}_D^{\mathbb{N}}$.

5. In this way all logical connectives have been essentially considered; in the remaining cases we replace $B \supset C$ with $\neg B \vee C$, $B \vee C$ with $\neg(\neg B \wedge \neg C)$, and $\forall x B(x)$ with $\neg \exists x \neg B(x)$. $\square$

The structure $\mathcal{R}$ can be seen particularly clearly within its ultrapower $\mathcal{R}_D^{\mathbb{N}}$. To each $a \in \mathbb{R}$ we can associate the sequence $\bar{a} = \{a_i\}_{i \in \mathbb{N}}/D$, where $a_i = a$ for all $i \in \mathbb{N}$. This mapping of $\mathcal{R}$ into $\mathcal{R}_D^{\mathbb{N}}$ is obviously well–defined and equally clearly a monomorphism. Indeed even more holds: if $A(a_1, \ldots, a_k)$ holds in $\mathcal{R}$, then $A(\bar{a_1}, \ldots, \bar{a_k})$ is true in $\mathcal{R}_D^{\mathbb{N}}$, and conversely. The mapping $a \mapsto \bar{a}$ is therefore called an *elementary embedding*. The elements of $\mathcal{R}$ the "ordinary" real numbers, are contained in the model $\mathcal{R}_D^{\mathbb{N}}$; we call them the standard elements of $\mathcal{R}_D^{\mathbb{N}}$, and all others non–standard elements.

Infinitely Small Elements of $\mathcal{R}_D^{\mathbb{N}}$

$\{1, \frac{1}{2}, \frac{1}{3}, \frac{1}{4}, \frac{1}{5}, \ldots\}/D = \epsilon$

(a) $\epsilon < (\overline{1/n})$ for all n, i.e.

$$\{i \in \mathbb{N} \mid \tfrac{1}{i} < \tfrac{1}{n}\} \in D \text{ for each fixed } n.$$

(b) $n \cdot \epsilon < 1$ for all n, i.e.

$$\{i \in \mathbb{N} \mid \tfrac{n}{i} < 1\} \in D \text{ for each fixed } n.$$

Infinitely Large Elements of $\mathcal{R}_D^{\mathbb{N}}$

$\{1, 2, 3, 4, \ldots\}/D = \omega$ 　　 $\omega > \bar{n}$ for all $n \in \mathbb{N}$.

For future brevity, we shall use the notation *$\mathcal{R}$, to denote the ultrapower $\mathcal{R}_D^{\mathbb{N}}$, and *$\mathbb{R}$ to refer to the set of its elements. Also, we shall omit the subscripts from the notation $0_D, 1_D, \leq_D, +_D$, etc.

To justify the definition of standard part, introduced in the first box on the next page, we have to show the existence and uniqueness of $^{\circ}a$.

Uniqueness. Let $x, y \in \mathbb{R}, x \neq y$, and suppose both $\bar{x} - a$ and $\bar{y} - a$ are infinitesimal. Then $(\bar{x} - a) - (\bar{y} - a) = \bar{x} - \bar{y} = x - y$ is infinitesimal; but $x - y \in \mathbb{R}$ and $|x - y| > 0$. So $x = y$.

Existence. If $a = \bar{x}$ for some $x \in \mathbb{R}$ we are done. Otherwise, we use Dedekind cuts to construct $^{\circ}a$ as follows. Define L, U by $L = \{x \in \mathbb{R} \mid \bar{x} < a\}$ and $U = \{x \in \mathbb{R} \mid \bar{x} > a\}$. As a Dedekind cut the pair (L, U) uniquely determines a real number $b \in \mathbb{R}$. We claim that $\bar{b} - a$ is infinitesimal. The number b must be either the greatest element of L or the smallest element of U. In the first case, then, $\bar{b} < a$. If $\bar{b} - a$ were not infinitesimal, then there would be some $e \in \mathbb{R}, e > 0$, such that $|\bar{b} - a| = a - \bar{b} \geq \bar{e}$, and therefore $\bar{b} + \bar{e} \leq a$ and $\bar{b} + \bar{e}/2 < a$. But then $b + e/2$ would be in L, contrary to our assumption that b is the greatest element of L. In the second case, if b is the least element of U, we can argue in exactly the same way.

Let $c \mapsto \overline{c}$ be the elementary embedding of $\mathcal{R}$ in $*\mathcal{R}$.

$a \in *\mathbb{R}$ is called infinitesimal if

$$a \neq 0 \text{ and } |a| < \overline{c} \text{ for all } c \in \mathbb{R}\, c > 0\,;$$

$a \in *\mathbb{R}$ is called infinite if

$$|a| > \overline{c} \text{ for all } c \in \mathbb{R}\,, c > 0\,;$$

$a \in *\mathbb{R}$ is called finite if

$$|a| < \overline{c} \text{ for some } c \in \mathbb{R}\,, c > 0\,.$$

In these definitions, $|a|$ is given by

$$|a| = \begin{cases} a & \text{if } a \geq 0 \\ -a & \text{otherwise} \end{cases}$$

For each $a \in *\mathbb{R}$ we define the standard part of a, denoted by $°a$, to be that element $x \in \mathbb{R}$ for which $(\overline{x} - a)$ is infinitesimal or zero.

We shall now use the ideas we have introduced to justify a modest part of analysis, and incidentally draw some parallels with high-school mathematics.

Limits, Continuity, and Derivatives

Limits. Let $\ell, a \in \mathbb{R}$ and let f be a function $*\mathbb{R} \mapsto *\mathbb{R}$. We set

$*\lim\limits_{x \mapsto a^-} f(x) = \overline{\ell}$ if and only if $°f(\overline{a} - y) = \ell$ for all infinitesimals $y > 0$.

$*\lim\limits_{x \mapsto a^+} f(x) = \overline{\ell}$ if and only if $°f(\overline{a} + y) = \ell$ for all infinitesimals $y > 0$.

$*\lim\limits_{x \mapsto a} f(x) = \overline{\ell}$ if and only if $°f(\overline{a} - y) = \ell$ for all infinitesimals y.

Continuity. Let $a \in \mathbb{R}$ and let f be a function defined on an open interval about $\overline{a}$.

f is *continuous at $\overline{a}$ if and only if $°f(\overline{a} + y) = °f(\overline{a})$ for all infinitesimal y.

Derivatives. Let $a, b \in \mathbb{R}$ and let f be defined on an open interval about $\overline{a}$.

$f'(\overline{a}) = \overline{b}$ if and only if $°(\frac{f(\overline{a}+y)-f(\overline{a})}{y}) = b$ for all infinitesimal y.

In the definition of derivative we can introduce the infinitesimal elements as differentials just as Leibniz did:

$$dx \quad \text{for} \quad y\,,$$

$$df \quad \text{for} \quad f(\overline{a} + dx) - f(\overline{a})\,;$$

Now the derivative $f'(a)$ can be written as the differential quotient

$$f'(a) = {}^{\circ}\!\left(\frac{df}{dx}\right).$$

Example

$$f(x) = x^2, \quad f'(x) = ?$$

$$df = (x + dx)^2 - x^2 = 2x\,dx + (dx)^2,$$

$${}^{\circ}\!\left(\frac{df}{dx}\right) = {}^{\circ}\!\left(\frac{2x\,dx + (dx)^2}{dx}\right) = {}^{\circ}(2x + dx) = 2x.$$

In the manner just indicated one can think through elementary differential calculus in an extremely elegant way. This is not only so for the manipulative aspects; even the proofs of the main results of differential calculus, for example Rolle's theorem, present themselves extremely intuitively. In all seriousness therefore one can ask oneself, whether one should not give a course in differential and integral calculus built up in this manner, that is taking the structure ${}^*\mathcal{R}$ instead of $\mathcal{R}$ as given. Here however it would not be a matter of constructing ${}^*\mathcal{R}$ as above; rather one would have to take ${}^*\mathcal{R}$ as given and start quasi-axiomatically from a few essential properties of ${}^*\mathcal{R}$, as one also does in the end for $\mathcal{R}$. {Such an attempt has been made by J. Keisler.}

However we want to leave such pedagogical questions packed away, and consider some uncompleted mathematical business: what have *limits, *continuity and *differentiation to do with the usual concepts?

Our definition of limits etc. involves the notion of the function f. Thus we fix once and for all the notion of a *defined* function, we say that f is definable if the relation $f(x) = y$ is definable by a formula $A(x, y)$ of elementary language, for which in $\mathcal{R}$ we have

$$\forall x\, \exists y\, A(x, y) \wedge \forall x\, \forall y_1\, \forall y_2 (A(x, y_1) \wedge A(x, y_2) . \supset y_1 = y_2).$$

This formula also holds in ${}^*\mathcal{R}$; in the extended structure $A(x, y)$ also defines a function, which for the sake of simplicity we also denote by f. Functions occurring in the definitions of *limits etc. are to be understood in this sense.

{A somewhat deeper reaching position would be the following: one forms an ultra power from the structure $\mathcal{A}$ of elementary analysis (see § 2) instead of from the structure $\mathcal{R}$. Then for each function f from $\mathcal{A}$ one has available an extended function $\overline{f}$ from ${}^*\!\mathcal{A}$ with exactly the same properties as are formulable in the language of $\mathcal{A}$. In a similar way one could also incorporate concepts of higher type into the construction, e.g. functionals; we refer to the wide–ranging and active literature on non–standard analysis, for example to A. Robinson.}

Now that the definitions of *limit etc. have been given their precise form, we examine the connection between the *concepts and the basic concepts coming from analysis. This is as simple as one could wish for.

Lemma 1 ${}^*\lim\limits_{x \mapsto a^-} f(x) = \bar{\ell}$ iff $\lim\limits_{x \mapsto a^-} f(x) = \ell$.

Proof. (a) Let $\lim_{x \mapsto a^-} f(x) = \ell$.

By the usual definition for each $\epsilon \in \mathbb{R}, \epsilon > 0$, there exists some $\delta \in \mathbb{R}, \delta > 0$, such that for all y $0 < y < \delta \supset |f(a - y) - \ell| < \epsilon$. Let $\epsilon > 0, \epsilon \in \mathbb{R}$ be chosen and a corresponding $\delta > 0$ determined. The statement $\forall y(0 < y < \delta \supset |f(a - y) - \ell| < \epsilon)$ holds in $\mathcal{R}$, is elementary and therefore holds in $^*\mathcal{R}$: $\forall y(0 < y < \bar{\delta} \supset |f(\bar{a} - y) - \bar{\ell}| < \bar{\epsilon})$. Now let y be infinitesimal, $0 < y$. Then in $^*\mathcal{R}$ $0 < y < \bar{\delta}$, hence also $|f(\bar{a} - y) - \bar{\ell}| < \bar{\epsilon}$. Since this argument applies for each positive ϵ in $\mathbb{R}$, by definition the magnitude of $|f(\bar{a} - y) - \bar{\ell}|$ is infinitesimal. Therefore $^\circ f(\bar{a} - y) = \ell$, and $^* \lim_{x \mapsto a^-} f(x) = \bar{\ell}$.

(b) Conversely let $^\circ f(\bar{a} - y) = \ell$ for all infinitesimals $y > 0$. We can choose some infinitesimal element $\delta > 0$. Then for arbitrary $\epsilon > 0, \epsilon \in \mathbb{R}$ and all $y, 0 < y < \delta$, we have that $|f(\bar{a} - y) - \bar{\ell}| < \bar{\epsilon}$, since each such y is itself infinitesimal. Formally expressed this says that, for each $\epsilon \in \mathbb{R}, \epsilon > 0$ the statements $\forall y(0 < y < \delta \supset |f(\bar{a} - y) - \bar{\ell}| < \bar{\epsilon})$ and $\exists \delta > 0 \forall y(0 < y < \delta \supset |f(\bar{a} - y) - \bar{\ell}| < \bar{\epsilon})$ are valid. For elementary reasons they also hold in $\mathcal{R}$ itself: $\exists \delta > 0 \forall y(0 < y < \delta \supset |f(a - y) - \ell| < \epsilon)$. Since $\epsilon > 0$ can be arbitrarily chosen in $\mathbb{R}$, the statement above reads as the definition $\lim_{x \mapsto a^-} f(x) = \ell$.

Lemma 2 f is **continuous at* $\bar{a}$ *if and only if* f *is continuous at* a.

Lemma 3 f *has the* **derivative* $\bar{b}$ *at* $\bar{a}$ *if and only if* $f'(a) = b$ *in the usual sense.*

The proofs of these lemmas are immediate applications of Lemma 1.

Further Reading

Leibniz, G.W.: A new method for Maxima and Minima as well as Tangents which is neither by Fractional nor by Irrational Quantities, and A Remarkable Type of Calculus for this, (1646-1716), English translation in: D.J. Struik: A Source Book in Mathematics, 1200-1800, pp. 272-280. Cambridge, Mass., Harvard University Press, 1969

Euler, L.: Institutiones calculi differentialis, Opera Omnia, Ser. I, vol. X, pp. 69-72, St. Petersburg 1755, excepts translated in: D.J. Struik: A Source Book in Mathematics, 1200-1800, pp. 384-386. Cambridge, Mass., Harvard University Press, 1969

Berkeley, G.: The Analyst, or a Discourse Addressed to an Infidel Mathematician, excepts in: D.J. Struik: A Source Book in Mathematics, 1200-1800, pp. 333-338. Cambridge, Mass., Harvard University Press, 1969

Skolem, Th.: Ueber die Nicht–Charakterisierbarkeit der Zahlenreihe mittels endlich oder abzählbarer unendlich vieler Aussagen mit ausschliesslich Zahlenvariablen, Fundamenta Mathematicae, vol. 23, pp. 150-161, (1934)

Łos, J.: Quelques Remarques, Théorèmes sur les Classes Définissables d'Algèbres, in: Mathematical Interpretation of Formal Systems, pp. 98-113, Amsterdam, North–Holland, 1954

Robinson, A.: Non Standard Analysis, Koninklijke Nederlandse Akademie van Wetenschappen Proceedings, Series A, vol. 64, pp. 432-440, (1961)

Keisler, H.J.: Elementary Calculus, Boston, Prindle, Weber & Schmidt Inc., 1976

§5 Axiom of Choice and Continuum Hypothesis

What are the real numbers? At least the question has become somewhat clearer
since it was posed in Section 1: any satisfactory answer must provide a frame
of reference for mathematical activity, in particular for proving theorems in
analysis. There seem to be two aspects of this activity that go hand in hand: on
the one hand there is what active mathematicians call "intuition" (without, if
they are wise, going into its psychological details), or "thinking in concepts", on
the other the mathematical formalism which, with the help of symbolic logic,
can be refined into a precision tool.

Commitment to a formal framework like the one in Section 3 restricts the
scope of possible mathematical assertions – albeit a restriction which in principle allows us to consider all pertinent questions as settled. To what extent
does our "intuition" support this restriction? What is the purchase price for
the unequivocal clarity, which the formalism provides? Evidently the reduced
expressive power of the language: the notions of limit and differential quotient,
as we remarked, have to be understood as schemata and must be introduced
specially for each function (and only definable functions are admissible). This
is not a mere inconvenience of formalization: there are some perfectly legitimate, even indispensable things which the formalization cannot capture. The
transcendental number π cannot be defined by a formula $A(x)$, because every
elementarily defined element is algebraic. Similarly for the sine function; every
definable function is made up of a finite number of pieces of algebraic functions
and therefore has only finitely many zeroes (unless it is the constant function
0). Even the natural numbers themselves escape definition by a formula $A(n)$
for the same reason.

This weakness in expressive power is what gives mathematicians the freedom
to invent models for the theory. Most of us in some way think of $^*\mathcal{R}$, but quite
legitimately work in $\mathcal{R}$ (which is perhaps what the early analysts did). The
mathematical results of the theory remain the same – they are just rather
meagre.

In Section 2 we already suggested one way to step beyond the limits of
the formal framework, by introducing functions as new entities. What does this
extension permit us to express? Just what we intend, or are there once again
non–standard models? Does the formalization decide all pertinent questions
expressible in the extended language, and, if not, are there any interesting open
problems? And what should one do about such problems if they exist?

The Natural Numbers

Let $N(y)$ be an abbreviation for the following formula:

$$\forall f[f(0) = 0 \wedge \forall x(0 \le x \wedge f(x) = 0 . \supset f(x+1) = 0) . \supset f(y) = 0.]$$

Peano's Axiom

(i) $N(0)$

(ii) $N(x) \supset N(x+1)$

(iii) $N(x) \wedge N(y) \wedge x + 1 = y + 1 . \supset x = y$

(iv) $N(x) \supset 0 \ne x + 1$

(v) $\forall x(A(x) \supset N(x)) \wedge A(0) \wedge \forall x(A(x) \supset A(x+1)) . \supset \forall y(N(y) \supset A(y))$

The Peano Axioms are demonstrable in elementary analysis. This is clear for (i)–(iii). To prove (iv) observe that $x < 0 \supset . \neg N(x)$, since the function

$$f(x) = \begin{cases} 0 & \text{for } x \ge 0 \\ 1 & \text{for } x < 0 \end{cases}$$

is definable, satisfies the premise of the definition of $N(y)$, but has the value $f(x) \ne 0$ for $x < 0$.

Proof of the Induction Axiom (v). Let $A(x)$ be a formula of elementary analysis and define the function f_A as follows:

$$f_A(x) = \begin{cases} 0 & \text{if } A(x) \\ 1 & \text{if } \neg A(x) \end{cases}$$

The existence of this function follows in turn from the axioms of elementary analysis. Start from the assumption of (v):

$$\forall x(A(x) \supset N(x)) \wedge A(0) \wedge \forall x(A(x) \supset A(x+1)).$$

Then we see that $f_A(0) = 0$, because $A(0)$ and

$$\forall x(0 \le x \wedge f_A(x) = 0 . \supset f_A(x+1) = 0),$$

since $\forall x(A(x) \supset A(x+1))$. From the definition of $N(y)$, it follows that

$$\forall y(N(y) \supset f_A(y) = 0)$$

and therefore $\forall y(N(y) \supset A(y))$.

Functions of Several Variables

We extend the language by adding a fixed two-argument function symbol p and the following axioms.

Pairing Axiom

$$[N(x) \wedge N(y). \supset N(p(x,y))] \wedge$$
$$[\forall x_1 \forall y_1 \forall x_2 \forall y_2(p(x_1,y_1) = p(x_2,y_2) \equiv .x_1 = x_2 \wedge y_1 = y_2)].$$

This permits us to introduce notation for functions of several variables:

$$f(x_1, x_2, x_3) \text{ for } f(p(x_1, p(x_2, x_3))), \text{etc.}$$

Comprehension Theorem

$$\forall x_1 \ldots x_n \exists! y A(x_1, \ldots, x_n, y) \supset$$
$$\exists f \forall x_1 \ldots x_n A(x_1, \ldots, x_n, f(x_1, \ldots, x_n)).$$

Recursion Schema

$$\forall g \forall h(\forall x_1 \ldots x_{n+1}[N(x_1) \wedge \ldots \wedge N(x_{n+1}). \supset . N(g(x_1, \ldots, x_n))$$
$$\wedge N(h(x_1, \ldots, x_{n+1}))]$$
$$\supset \exists f \forall x_1 \ldots x_{n+1}[N(x_1) \wedge N(x_2) \wedge \ldots \wedge N(x_{n+1}).$$
$$\supset f(x_1, \ldots, x_n, 0) = g(x_1, \ldots, x_n)$$
$$\wedge f(x_1, \ldots, x_n, x_{n+1} + 1) = h(x_1, \ldots, x_n, f(x_1, \ldots, x_{n+1})).$$

We leave it to the reader to carry out the proofs of the Comprehension Theorem and Recursion Schema. They have been formulated here mainly to indicate how wide areas of analysis and number theory can be justified with the formal framework we have presented. This is not the place to carry out the details.

The results that we have derived about analysis are already sufficient for the informed logician to see that our axiomatization, if it is consistent, cannot possibly be complete, and cannot be effectively extended to a complete theory. (Following Gödel this phenomenon already enters with the theory of natural numbers.) Somewhat less hidden than in number theory (where such assertions have only been discovered in recent years) there are quite natural, and even central questions of elementary analysis that cannot be settled by appeal to the given axioms alone. Among these are the Axioms of Choice and Continuum Hypothesis.

One criterion for acceptance of an independent principle, like the Axiom of Choice, is based on mathematicians' (somewhat disputed) intuitions; another comes from examining the mathematical consequences of the principle. In the latter case the attractive, simplifying consequences must be weighted against those which appear paradoxical, like for example the following.

Banach–Tarski Paradox

As refined by R. Robinson, the paradox goes like this:

The solid unit ball S can be partitioned into six disjoint subsets

$$S = A_1 \cup A_2 \cup A_3 \cup A_4 \cup \{0\} \cup \{P\}$$

in such a way that the whole ball S can be built up by rotations from the two sets
$$A_1 \text{ and } A_3 \cup \{0\}$$
and can also be reconstructed using rotations and translations from the three sets
$$A_2, A_4 \text{ and } \{P\}.$$

Thus appropriate partitions and congruence can be used to transform one ball into two balls of the original size!

In spite of this and other paradoxical consequences of the Axiom of Choice, mathematicians (with some exceptions) have decided not to dispense with the axiom. The paradoxes are explained away as shortcomings of our naive concepts, for example our notion of "volume" as contrasted with a formally defined concept of volume. Such shortcomings are not unfamiliar: just think of the examples of continuous but nowhere differentiable functions and of how analysis is enhanced by such seemingly paradoxical counterexamples.

Cantor, the inventor of set theory struggled with the question of the size of the continuum. It is typical of the pioneer's situation that he does not have the benefit of others' experience to help him distinguish hard problems from easy ones. Cantor's conjecture was that the cardinality of the continuum is the least uncountable cardinal. Like the Axiom of Choice, the Contiuum Hypothesis is consistent with (Gödel) but independent of (Cohen) the usual axiomatizations of set theory. The difference is that, until now, few have felt there is sufficient reason to extend the accepted framework of mathematics to include either the Continuum Hypothesis or its negation. Unlike the Axiom of Choice, if the Continuum Hypothesis is used in a proof it conventionally must always be explicitly mentioned as an additional assumption.

The Continuum Hypothesis for the structure $\mathcal{A}$ of elementary analysis can be stated very easily: it says simply that every subset X of $\mathbb{R}$ has cardinality which is either less than or equal to $\aleph_0$ or else greater than or equal to the cardinality of $\mathbb{R}$. In the first case there is a function f mapping $\mathbb{N}$ onto X, in the second a function g that maps X onto $\mathbb{R}$. If we take X to be the set defined by the relation $h(x) = 0$, this can be expressed as follows.

Continuum Hypothesis

$$\forall h [\exists f \forall y (h(y) = 0 \supset \exists x (N(x) \wedge y = f(x)))$$
$$\vee \exists g \forall y \exists x (h(x) = 0 \wedge y = g(x))]$$

Now that we are able to formulate the Axiom of Choice and Continuum Hypothesis for elementary analysis, let us conclude with Gödel's and Cohen's theorems about them.

Gödels Consistency Theorem *Any formal proof of a contradiction that arises in elementary analysis with the Axiom of Choice and Continuum Hypothesis added to the axioms can be effectively transformed into a derivation of a contradiction from the original axioms of elementary analysis alone.*

Cohen's Independence Theorem *Any formal proof of a contradiction that arises in elementary analysis together with the Axiom of Choice and the negation of the Continuum Hypothesis can be effectively transformed into a proof of a contradiction from the original axioms alone.*

Further Reading

Hausdorff, F. : Grundzüge der Mengenlehre, (1914) pp. 399-403, 469-473, New York, Chelsea, (reprint 1949)

Robinson, R.M. : On the Decomposition of Spheres, Fundamenta Mathematicae, vol. 34, pp. 246-260, (1947)

Gödel, K.: What is Cantor's Continuum Problem?, in: P. Benacerraf and H. Putnam: Philosophy of Mathematics, pp. 258-273, Englewood Cliffs, Prentice-Hall, 1964

Gödel, K.: Consistency Proof for the Generalized Continuum–Hypothesis, Proc. Nat. Acad. Sci. USA, vol. 25, pp. 220-224, (1939)

Cohen, P.: The Independence of the Continuum Hypothesis I,II, Proc. Nat. Acad. Sci. USA, vol. 50, pp. 1143-1148, and vol. 51, pp. 105-110, (1963, 1964)

Scott, D.: A Proof of the Independence of the Continuum Hypothesis, Mathematical System Theory, vol. 1, pp. 89-111, (1967)

Chapter II. Geometry

§ 1 Space and Mathematics

In the question of the concept of space, even more evidently than in the question of the real numbers, the problem of the relation between mathematics and the so–called real world is posed. Newton formulated his position as follows: "Geometry is founded in practical mechanics, and indeed is no more than that part of mechanics as a whole, which originates in and is confirmed by the art of measurement." Or Gonseth: "Geometry is the physics of arbitrary space." The sense of geometry may therefore lie in finding a solid basis for the art of measurement (one that imposes a duty): mathematical consequences of axioms about space ought to be verifiable in actual surveying (hence comes the name of this science). Like physicists always we are then disposed to buy the ideal situation and to treat discrepancies as "incidental" and not "systematic" mistakes of measurement.

Newton and classical physics therefore begin with the implicit assumption of the existence of a substratum independent of physics, namely empty space, and form the concept of geometry by idealizing data such as points, lines, distances, angles and their relations. That there exists a wide consensus about the properties of these idealized data was already agreed by the Ancients; it survives unbroken in school geometry. In this sense Euclidean geometry forms a framework, similar to the framework of the continuum.

Euclidean geometry is given the duty of serving measurement; real numbers represent in its system of distances the universal system of measurement for magnitudes (Archimedean of course!). The foundational problem of geometry is seen in this light by different geometers and so is comparable with the foundational problem of real algebra and analysis considered in Chapter I. In the following sections we wish to present and discuss an axiomatization programme resulting from this position, maintaining the critical attitude and experience inherited from Chapter I towards the exchange relation between the expressibility of languages and incompleteness of axiom systems, but shorn of definitional helplessness, as exhibited by Euclid's "a point is something which has no parts".

Here is the place to speak briefly about the basic questionability of the programme just sketched. First from the point of view of *physics*: for physical results is it necessary to postulate an independent preexisting geometric substratum, i.e. is so–called "space" a physically necessary hypothesis? Or rather should one not integrate measurement in some way into physics, possibly in such a way that it cannot be later detached as a separate subconcept of purely "geometric" character. Modern physics brings us close to this position. – Even if one takes the concept of a geometric substratum of physics as hypothesis, it is *geometrically* questionable whether one ought to assume that this is classical, Euclidean, elementary geometry. This is the methodological position of the so–called space problem, which has occupied the geometry of the last hundred years in an essential way, and is the starting point for the development of differential geometry. Here one has only to name Riemann, Helmholtz and Lie. We refer to the extremely stimulating accounts of Riemann ("On the hypotheses which underpin geometry") and Helmholtz ("On the facts which underpin geometry"). The collection of Freudenthal gives an overview of the history of the posing of this question by Lie, Weyl etc. The problem is still alive today: what follows for the structure of space, if one presupposes free movability of finite (infinitesimal) rigid bodies? One can also pose the space problem in topological rather than in differential geometric terms – what are the topological properties of a topological space, which suffice to characterize Euclidean space? Are there such which in some sense can be called "obvious"? This problem appears to be very difficult – for example one can consult the work of Borsuk. Here we do not wish to recount once more the history of non-Euclidean geometry, of Gauß and the (not sounded) Boeotian protest, of Nikolai Ivanovich, of the son of Herrn Bolyai; this is easy to find elsewhere. Instead we finish this section – somewhat anecdotally – with a few references to the initial difficulties of modern axiomatics. Hilbert explicitly took the position that the axioms implicitly define the basic concepts contained in them. Nonsense said the "Boeotians", if one inverts in the unit circle, the axioms (of incidence) remain true, but a line is no longer a line, but a circle, how then can "line" be defined? Frege produced a more subtle criticism: an axiom system is something like a system of equations which one cannot solve. If we want to answer the question, whether something or other is a point, say, my watch, we already have problems with the first axiom: This refers to two points. Frege's parody of Hilbert's axiomatization is telling: "Statement: we assume entities, which we call Gods.

Axiom 1: Each God is all powerful.

Axiom 2: There exists at least one God."

This parody hits the central point of the axiomatization: With Hilbert's "we assume entities...", geometry becomes pure mathematics.

Further Reading

Riemann, B.: Ueber die Hypothesen, welche der Geometrie zu Grunde liegen (1854). Neu herausgegeben und erläutert von H. Weyl, in: Das Kontinuum und 3 Monographien, New York, Chelsea Publishing, Company.

Von Helmholtz, H.: Ueber die Thatsachen, die der Geometrie zum Grunde liegen (1868), in: Wissensch. Abhandlungen, vol. 2, pp. 618-639, Leipzig, Barth, 1883

Freudenthal, H.: Im Umkreis der sogenannten Raumprobleme, in: Bar–Hillel et al.: Essays on the Foundations of Mathematics, pp. 322-327, Amsterdam, North–Holland, 1962

Borsuk, K.: Grundlagen der Geometrie vom Standpunkte der allgemeinen Topologie aus, in: Henkin, Suppes & Tarski: The Axiomatic Method, with Special Reference to Geometry and Physics, pp. 174-187, Amsterdam, North–Holland, 1959

§ 2 Axiomatization by Means of Coordinates

Since we have imposed on Euclidean Geometry the duty of using the field $\mathcal{R}$ of real numbers as distance system, this is easiest to understand as a twodimensional vector space $\mathcal{E}$ over $\mathbb{R}$. With the help of the scalar product the distance function is defined to be

$$\|x,y\| = \sqrt{(x-y)\cdot(x-y)} \text{ for arbitrary } x,y \in \mathcal{E}.$$

The fundamental geometric concepts: point, line, circle, angle, congruence etc. are now to be defined as concepts for the vector spaces $\mathcal{E}$. Then the essential properties of these concepts and the relations between them must be organized into a system of geometric entities and axioms, so that in their totality they again characterize the vector space $\mathcal{E}$. This, very vaguely expressed, is the programme for axiomatizing elementary geometry, that we have before us. Which concepts and relations are to be considered as "geometric" for $\mathcal{E}$? We agree that they are to be those which are preserved under isometries, that is under distance preserving maps from $\mathcal{E}$ to itself. For example the image of a circle under a map $f : \mathcal{E} \to \mathcal{E}$ is still a circle, if f is an isometry, that is

$$\|x,y\| = \|f(x),f(y)\|$$

for all $x,y \in \mathcal{E}$.

Formally therefore we can define the fundamental geometric concepts as follows:

$R \subseteq \mathcal{E}^n$ is an n–ary *geometric relation* iff for all isometries f of $\mathcal{E}$ one has

$$\langle x_1,\ldots,x_n\rangle \in R \equiv \langle f(x_1),\ldots,f(x_n)\rangle \in R.$$

A *class* C of subsets of $\mathcal{E}$ is called *geometric*, if for all isometries f and all $\alpha \in C$ one has $\{f(x) : x \in \alpha\} \in C$.

In this sense the usual basic concepts of elementary geometry, such as lines, circles, triangles etc. are geometric classes, and the concepts of incidence and congruence are geometric relations (on points).

Our axiomatization programme can be yet more sharply formulated as follows: one can choose a finite family of geometric classes and geometric relations on these classes and look for a system of axioms for this system of elements and relations so that: if an elementary theorem holds for the basic concepts as defined in $\mathcal{E}$, then this theorem is provable in the axiom system and conversely. (Here, as earlier, we call a statement elementary if it is expressible in the first order language with these relations and elements.) Our procedure for finding such an axiom system consists in the following three tasks: (1) The points of a linear (1–dimensional) subspace of $\mathcal{E}$ uniquely determine the elements of the ground field of $\mathcal{E}$, namely the real numbers. By means of simple geometric constructions, such as the laying down of linear segments and proportionality theorems, the field operations can then be carried out. As a next step, it is easy to use geometric concepts and their properties to formally define these constructions and to make the properties of the field operations formally provable. (2) The field structure introduced in this way on a line will then be used to coordinatize the whole Euclidean plane, and one shows that the vector space obtained in this way is isomorphic to the original vector space $\mathcal{E}$. (3) Exact proof analyses then finally give the applied properties of the geometric relations and hence the axiomatization.

We wish to carry out this axiomatization programme in such a way, that essentially we obtain the known Hilbert axiom system (for the Euclidean plane). This choice is of course completely arbitrary; it is based on historic interest and the accessibility of the literature.

So let us begin by choosing basic geometric elements

> *Lines:*   variables denoted by $x, y, z \ldots, g, \ell, \ldots, a, b, c, \ldots$
> *Points:*   variables denoted by $X, Y, Z, \ldots, P, Q, \ldots, A, B, C, \ldots$,

and by first only taking the (geometric) relation of *incidence* into the vocabulary. Denote this by $P \in \ell$. We take over the numbering (but not the formal expression) of the Hilbert axioms.

Incidence and Parallel Axioms

I_1 & I_2 : $A \neq B \supset \exists! \, a (A \in a \land B \in a)$;

I_3 : $\forall a \exists A \exists B (A \neq B \land A \in a \land B \in a) \land \exists A \exists B \exists C$
$\neg \exists a (A \in a \land B \in a \land C \in a)$;

Definition $a \parallel b \; :\equiv \; \exists A (A \in a \land A \in b) \supset a = b$

IV^* : $\forall a \forall A (\neg A \in a \supset \exists! \, b (A \in b \land a \parallel b))$.

From the axioms I_1–I_3 the existence of the basic figure clearly follows. It will be of central importance in what follows. It consists of two lines g and g' meeting each other at the intersection point O and containing points E, E'

distinct from O. Next we want to introduce geometrically the field operations on the line g, that is, define them by means of the incidence relations alone, with the point E representing the multiplicative identity, and the point O the neutral element for addition. {In what follows we use the usual geometric manner of speaking: cut, lies on, goes through, etc. without further definition. In the diagrams we use marks on lines in order to indicate sets of parallel lines. $g \times h$ indicates intersection.}

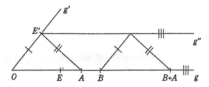

Fig. 2.1. Addition

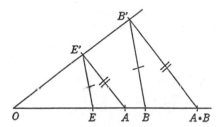

Fig. 2.2. Multiplication

The possibility of drawing figures as above for the construction of addition and multiplication rests on the following lemma, which can be easily proved using the axioms introduced so far.

Lemma *In the case that $h \neq g$ and $h \times g$, it follows from $h' \parallel h$, that $h' \times g$.*

The question as to whether addition and multiplication are well–defined, as well as the question of the validity of the field axioms, impose conditions on the validity of certain closure theorem. The most important among them is Desargue's Theorem, which we include initially among the axioms.

Desargues' Theorem

D :   *If corresponding sides of two triangles are parallel, the lines joining corresponding vertices either meet at a point or are parallel. Conversely, if two triangles are so positioned that the lines joining corresponding vertices are either parallel or meet at a point, and if in addition two pairs of corresponding sides of the triangles are parallel, then the third pair of sides of the two triangles is also parallel.*

A useful remark follows immediately from Desargues' Theorem in the definition of $B + A$ one can draw the line g'' through any point of g' other than O, instead of through E'.

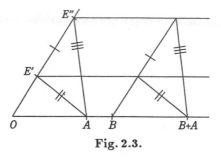

Fig. 2.3.

The following proofs for the field axioms are mostly given by repeated application of the Desargues Theorem, and we make this clear by suitable shading of triangles which are in perspective with one another.

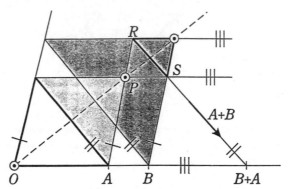

Fig. 2.4. Commutativity of Addition

Because of D for the shaded triangles $\triangle$ the points $\odot$ lie on a line. Given the perspective position of the outlined triangles $\triangle$ with centroid at P, the line RS passes through $B + A$.

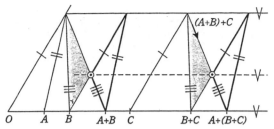

Fig. 2.5. Associativity of Addition

Because of D for $\triangle$ the lines with a " $\vee$ " are parallel. Because of D for the shaded triangles $\triangle$ it follws that $A+(B+C)=(A+B)+C$.

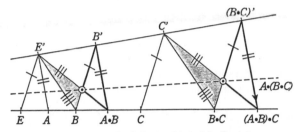

Fig. 2.6. Associativity of Multiplication

In order to prove the commutativity of multiplication we need a further closure theorem, which we again postulate as an additional axiom. This is the theorem of Pappus and Pascal, usually referred to in short as Pascal's theorem.

Pascal's Theorem P : *Let the points* $1, 2, 3$ *(respectively* $1', 2', 3'$ *) belonging to two lines be such that the line segments* $12'$ *and* $1'2$ *as well as* $23'$ *and* $2'3$ *form parallel pairs. Then the line segments* $13'$ *and* $1'3$ *are also parallel.*

Desargue's Theorem follows from that of Pascal; the complete proof, which we shall spare ourselves here, consists of several special cases.

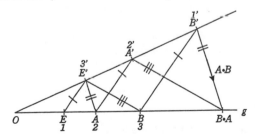

Fig. 2.7. Commutativity of Multiplication

We label the points on g and g' as shown and use P to deduce that $A \cdot B = B \cdot A$.

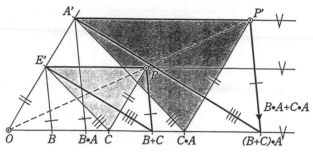

Fig. 2.8. Distributive Law

We apply D twice and so prove distributivity.

The existence of zero (namely O), one (namely E), of negative and inverse elements is a trivial exercise – hence finally we have at least proved the field axioms. In order to introduce the relation of order in the field we must first look for a *geometric* concept; a natural choice is the relation

$$ABC \; : \quad \text{``}B \text{ lies between } A \text{ and } C \text{ and is distinct} \\ \text{from both of them''}.$$

This relation immediately serves to introduce the relation of order in the constructed field. We base this on the more simply defined property of positivity.

Definition

$$A > O :\equiv A \neq O \wedge \neg AOE$$

By means of the concept of positivity we formulate the order axioms of the field as

$$\neg(O > O) \,; \quad A > O \vee -A > O \vee A = O \,;$$
$$A > O \wedge B > O. \; \supset . \, A + B > O \wedge A \cdot B > O.$$

From suitably chosen axioms these must be derived through the relation of "betweenness". Following Hilbert these read as follows

---

Axioms of Betweenness

$\mathrm{II}_1 :$ $ABC \supset .CBA \wedge \exists a(A \in a \wedge B \in a \wedge C \in a) \wedge A \neq B \wedge A \neq C \wedge B \neq C \,;$

$\mathrm{II}_2 :$ $A \neq B \supset \exists C \; ABC \,;$

$\mathrm{II}_3 :$ $ABC \supset \neg ACB \wedge \neg BAC \,;$

$\mathrm{II}_4 :$ $\neg \exists x (A \in x \wedge B \in x \wedge C \in x) \wedge A \notin a \wedge B \notin a \wedge C \notin a$
    $\wedge D \in a \wedge ADB . \supset . \exists E(E \in a \wedge AEC) \vee \exists F(F \in a \wedge BFC).$
    (Axiom von Pasch)

An exact derivation of the order axioms from the betweenness axioms is relatively difficult, since the axioms II are taken as just adequate. Since such a demonstration is of no further interest to us, we pass over it (one can consult Hilbert's book), and formulate, as last axiom, the one which makes our constructed field isomorphic to the real numbers, namely the completeness axiom.

Completeness Axiom

V: Let M and N be two non-empty sets of points on a line g, and suppose there exists a point C such that for all $A \in M$, $B \in N$ we have CAB. Then there exists a point D such that for all $A \in M, B \in N$ distinct from D we have ADB.

It follows immediately from V that the field constructed on g is complete. And in this way task (1) of our axiomatization programme is concluded.

In order to complete task (2) we use the field constructed on the line g to coordinatize the whole plane. For this we must first of all construct a field on the line g'; we take this as having been done in the same way as for g, by exchanging E and E'. It then follows that the fields on g and g' are isomorphic, the isomorphism being given by projection parallel to the line EE'.

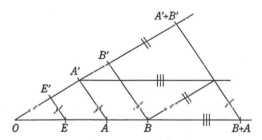

Fig. 2.9. Addition

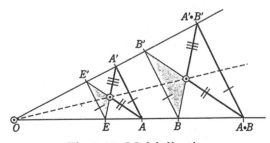

Fig. 2.10. Multiplication

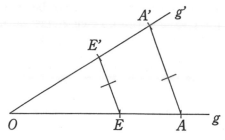

Fig. 2.11. Order

$OEA \equiv OE'A'$ because of the axioms in class II, in particular Pasch's axiom II_4.

With each point we can now associate two coordinates, real numbers X, Y, by means of the following definition:

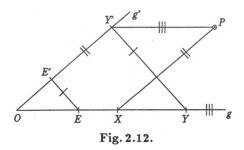

Fig. 2.12.

Finally we make the plane into a vector space $\mathcal{V}$ by means of the following definition of the inner product of two points P and Q: if X_1, Y_1 (resp. X_2, Y_2) are the coordinates of P (resp. Q), then $P \cdot Q := X_1 \cdot X_2 + Y_1 \cdot Y_2$, where addition and multiplication are as defined on the coordinate axes. However there still remains an arbitrary element in our coordinatization – we were completely free in the choice of coordinate axes and unit points. Mathematically this implies that the map $(X, Y) \mapsto (\overline{X}, \overline{Y})$ of the plane onto itself described by means of coordinates is in general no isometry. (Here X, Y denote the coordinates of a point P with respect to the axes g and g', $\overline{X}, \overline{Y}$ the coordinates of the same point with respect to the axes $\overline{g}, \overline{g}'$.) What is missing is the ability to express that the coordinate axes are perpendicular to each other, and that the distances OE and OE' are equal. To this end we introduce the congruence of line segments

$AB \approx CD$ "the segment AB is congruent to the segment CD",

and the congruence of angles

$ABC \approx EFG$ "the angle at B is congruent to the angle at F".

For these new basic geometric concepts we formulate the axioms of Hilbert's third group as follows:

Congruence Axioms

III$_1$:　$A \neq B \wedge C \neq D. \supset \exists E(CDE \wedge AB \approx DE)$;

III$_2$:　$AB \approx EF \wedge CD \approx EF. \supset AB \approx CD$;

III$_3$:　$AB \approx A'B' \wedge BC \approx B'C' \wedge ABC \wedge A'B'C'. \supset AC \approx A'C'$;

III$_4$:　$\neg \exists a(A \in a \wedge B \in a \wedge C \in a) \wedge \neg \exists a(A' \in a \wedge B' \in a \wedge C^* \in a). \supset$
$\exists C' \exists a[A' \in a \wedge B' \in a \wedge \neg \exists D(D \in a \wedge C^* DC') \wedge ABC \approx A'B'C'$
$\wedge \forall C''((\neg \exists E(E \in a \wedge C^* EC'') \wedge ABC \approx A'B'C'') \supset$
$(C'' = C' \vee B'C''C' \vee B'C'C''))]$;

III$_5$:　$AB \approx A'B' \wedge AC \approx A'C' \wedge BAC \approx B'A'C'. \supset ABC \approx A'B'C'$;

* :　$AB \approx CD \supset C \neq D$;

* :　$A \neq B \supset AB \approx BA$;

* :　$ABC \approx A'B'C' \supset \neg \exists a(A' \in a \wedge B' \in a \wedge C' \in a)$;

* :　$A_1B_1C_1 \approx A_3B_3C_3 \wedge A_2B_2C_2 \approx A_3B_3C_3. \supset A_1B_1C_1 \approx A_2B_2C_2$;

* :　$\neg \exists a(A \in a \wedge B \in a \wedge C \in a) \supset ABC \approx CBA$;

* :　$\neg \exists a(A \in a \wedge B \in a \wedge C \in a) \wedge (BA'A \vee BAA'). \supset ABC \approx A'BC$.

The axioms marked * are missing in Hilbert; they become necessary because of the way in which the concept (angle congruence as a 6-ary relation between points) is formulated. One could now use this fully-fashioned axiom system of Hilbert as the basis for the strict derivation of the results of elementary Euclidean geometry, for example the congruence theorems for triangles, Pythagoras' Theorem and so on. We do not wish to dwell on this but simply remark:

Suppose that in the Euclidean plane $\mathcal{E}$ there are two lines g, g' meeting at a point O, and points E, E' on g, g', respectively, so chosen that g and g' are perpendicular and $OE \approx OE'$, then the vector space $\mathcal{V}$ constructed as above from points of the plane is isomorphic to $\mathcal{E}$, the two-dimensional vector space over $\mathbb{R}$.

Further Reading

Artin, E.: Geometric Algebra, Chap. II, pp. 51 ff., New York, Interscience, 1957

Veblen, O. & Young, J.W.: Projective Geometry, vol. I, Chap. VI, pp. 141-168, New York, Blaisdell.

Hilbert, D.: Grundlagen der Geometrie, 7. edition, § 24 - § 27 and § 32, Stuttgart, Teubner, 1930

Schwabhäuser, W.: Ueber die Vollständigkeit der elementaren Euklidischen Geometrie, Zeitschrift für math. Logik und Grundlagen der Mathematik, vol. 2, pp. 137–165, (1956)

§3 Metatheoretical Questions and Methods in Elementary Geometry

The completeness axiom V for plane Euclidean Geometry leaves us facing the problem already considered in Chapter I: In what sense are the sets M and N named in it to be understood? In other words, how to adequately express the content of Axiom V in our formal language? And once more we choose the same solution – we identify sets with extensions of predicates, in the present case with predicates in the first order language of elementary geometry built the basic concepts introduced in §2. The resulting completeness axiom, made elementary, is called the Tarski Schema.

V_T :   For each pair of formulas $H_1(.)$ und $H_2(.)$ we have:
$\exists C \forall A \forall B(H_1(A) \wedge H_2(B) . \supset ABC)$
$\supset \exists D \forall A \forall B(A \neq D \wedge B \neq D \wedge H_1(A) \wedge H_2(B) . \supset ADB)$.

It immediately follows from V_T that the ordered field constructed on g satisfies the (elementary) completeness axiom for the theory of real algebra (Chap. I, §2). Thanks to our construction in §2 the completeness and decidability of this latter theory is transferred to elementary Euclidean geometry. Given this we want to exhibit a parallelism between geometry and linear algebra, which makes clear, also in a formal sense, the relations resulting from coordinatization. To this end we choose a formulation of the respective languages, which allows us to see clearly the correspondence between fundamental concepts of geometry and linear algebra. In most cases we spare ourselves the full writing out of the definitions.

Notice that each formula in the language of geometry, as expanded by definitions, can be expressed as a formula in the language of algebra, as expanded by definitions. Therefore the distinction between geometry and algebra lies more in the display of those facts which we wish to consider as axioms. For the sake of an overall picture we set this out once again as follows.

Language of Geometry		Language of Linear Algebra	
$P,Q,R...$	Point variables	$P,Q,R...$	vector variables
$O,E,E'...$	singled–out points	0	zero vector
$\ell,m,n...$	line variables	E,E'	orthonormal–basis
		ℓ,m,n	linear subspace variables
g,g'	singled–out lines	g,g'	linear subspaces spanned by E,E'
$\in$	relation of incidence	$\in$	relation of belonging
$\approx$	segment congruence	$AB \approx CD :\equiv (A-B)\cdot(A-B) = (C-D)\cdot(C-D)$	
PQR	"betweenness" relation	$PQR :\equiv ...$	
$\approxeq$	angle congruence	$PQR \approxeq UVW \quad :\equiv ...$	
x,y,z	variables for points on g	x,y,z	variables for elements of the ground field
$P\cdot Q = x$	$:\equiv ...$	$P\cdot Q$	inner product
$x+y = z$	$:\equiv ...$	$+$	field addition
$-y = z$	$:\equiv ...$	$-$	field subtraction
$x\cdot y = z$	$:\equiv ...$	$\cdot$	field multiplication
$y^{-1} = z$	$:\equiv ...$	$^{-1}$	field inversion
$x = 0$	$:\equiv x = O$	0	zero
$x = 1$	$:\equiv x = E$	1	one
$x>0$	$:\equiv ...$	$<$	positivity

Axioms of Geometry: Γ	Axioms of Linear Algebra: Λ
I,II,III,IV*,V$_T$,D,P, together with $(g \perp g' \wedge OE \approx OE' \wedge O \in g \wedge$ $O \in g' \wedge E \subset g \wedge E' \in g')$	Axioms for real closed fields (for variables $x, y, \ldots$) together with the usual axioms for vector spaces, bases E, E'; incidence for linear subspaces.

Metatheorem

For each formula F for the expanded language we have: if F is provable Γ, then F is provable in Λ and conversely.

We have already accomplished the main steps of the proof of the metatheorem above in § 2, at least in one direction. The other is "analytic geometry"; we do not want to involve ourselves with this yet again. Instead we turn to the question of the completeness and decidability of the axiom system Γ for plane Euclidean geometry as above.

In Chapter I we have proved both these metamathematical properties in detail for the axiomatization of real closed fields. This result carries over in the simplest manner to the axiom system Λ sketched above for linear algebra; detailed demonstration is superfluous. However, as we will immediately show, from the completeness and decidability of Γ follows

> ### Completeness and Decidability of Elementary Plane Euclidean Geometry
>
> *For each statement F in the language of elementary geometry either F or $\neg\, F$ can be proved from the axioms Γ; the question as to whether a given statement is provable is effectively decidable.*

For the *proof* we consider an arbitrary formula F from elementary geometry. We reexpress this formula in the language of linear algebra, as expanded by definitions. It then follows from the completeness and decidability of Λ that either F or $\neg F$ is provable in Λ, and that we can effectively decide which of these is the case. By the metatheorem the same holds for Γ.

With our procedure we have therefore found a complete axiom system for elementary plane Euclidean geometry. However, it is rather circuitous, and the question is raised as to whether the axiom system can be reduced. Certainly, individual axioms are avoidable; for example the Pascal and Desargues Theorems follow from the remaining axioms. However, we do not wish here to go into the amusing individual questions about the independence of different axioms and the resulting choice of non–Euclidean, non–Pasch geometries etc. This also for the following fundamental reason: The choice of the basic concepts and the axioms for geometry is clearly bound up with a great deal of arbitrariness and historical chance. If we now pick out individual axioms and take their negation instead (assuming there is no contradiction), then we possibly emphasize this arbitrary nature, and obtain an extremely complicated, even baroque, version of "geometry", which does not need to express some or other legitimate questions from the domain of real spatial problems.

Research into the bases of geometry already concerned itself early on with the question of choosing fundamental geometric concepts which, as simple and as low as possible, should be set out at the beginning.

First of all one easily sees that one can get away with one class of objects, for example with points; lines are introduced as pairs of points distinct from each other. Formally this expresses itself through a translation mechanisms which associates statements without line variables to equivalent statements with. Thus

$$\exists\, a F(a) \equiv \exists\, A_1 \exists\, A_2 (A_1 \neq A_2 \land F')$$

where F' results from $F(a)$ by replacing each atomic formula $B \in a$ by

$$(B = A_1 \lor B = A_2 \lor B A_1 A_2 \lor A_1 B A_2 \lor A_1 A_2 B),$$

and each $a = b$ by

$$(A_1 \in b \land A_2 \in b).$$

In a further step angle congruence can be set aside, for from the first congruence theorem for triangles follows

$$ABC \approx A^*B^*C^* . \equiv . \quad \exists A' \exists C'(A'B^* \approx AB \wedge A'C' \approx AC \wedge B^*C' \approx BC \wedge$$
$$(B^*A'A^* \vee A' = A^* \vee B^*A^*A') \wedge$$
$$(B^*C'C^* \vee C' = C^* \vee B^*C^*C')) .$$

Plane Euclidean geometry as above can therefore be set up with points alone, using only the relations of betweenness and of congruence for line segments. Tarski set out a corresponding axiom system (see the guide to further reading); he writes

$$\beta ABC \text{ for } ABC \vee A = B \vee B = C , \text{ and}$$

$$\delta ABCD \text{ for } AB \approx CD \vee (A = B \wedge C = D) .$$

It is easy to reduce the basic concepts still further, namely one notes that βABC can be defined by means of δ. In order to simplify the notation (we have already removed the different types of variable), from now on we use lower case letters as point variables. We define

$$xy \leq yz := \forall u[\delta yuuz \supset \exists v(\delta xvvy \wedge \delta vyyu)] ;$$

the geometric interpretation of $xy \leq yz$ is that the distance between x and y is smaller or equal to the distance between y and z. Namely

$$\|x,y\| \nleq \|y,z\| = \exists u(\delta yuuz \wedge \forall v(\delta vyyu \supset \neg \delta xvvy)) ,$$

as is made plain in the following diagram:

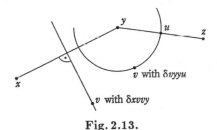

v with $\delta vyyu$

v with $\delta xvvy$

Fig. 2.13.

Next one represents the modified betweenness–relation β as

$$\beta xyz := \forall u(ux \leq xy \wedge uz \leq zy. \supset u = y) ,$$

which we again set out by means of a diagram (and a reference to school geometry):

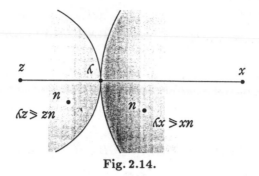

Fig. 2.14.

Therefore plane geometry might make do with only the 4–variable predicate δ. In order to reduce still further, one notes that in order to define β and $\le$ the predicate δ is used only in the form $\delta xyyz$, that is with equal middle variables. This is the relation which already in 1908 Pieri recognized as sufficient.

$$\pi xyz :\equiv \delta xyyz .$$

Indeed one can regain δ itself from π: one defines β and $\le$ as above (applying the symbol π), and then defines

$$coll \ xyz \quad :\equiv \quad \beta xyz \vee \beta yxz \vee \beta xzy ,$$
$$sym \ xyz \quad :\equiv \quad \forall u(coll \ xyu \wedge \pi xyu \equiv . u = x \vee u = z) .$$

The first of these expresses that x, y, z are collinear, the second that x and z are symmetrically placed with respect to y. Then it is clear that in the Euclidean plane there is a theorem which allows the following definition of δ:

$$\delta xyuv :\equiv \exists r \exists t(sym \ xru \wedge sym \ yrt \wedge \pi tuv) .$$

The following diagram illustrates the content of the theorem:

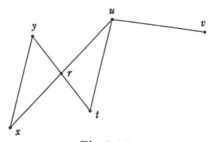

Fig. 2.15.

Further reductions were found by Bernays, Scott (1956) and Henkin (1962):

νxyz : the angle at y is a right angle,
$\pi' xyz$: the triangle xyz is equilateral,
$\nu' xyz$: the triangle xyz is right–angled.

How far can one go? More particularly, with what means can one prove that certain defined relations or types of relation do *not* suffice as a basis for geometry? Here is an example:

Theorem *In the Euclidean plane it is not possible to define one or more binary operations on points from π in an elementary way, so that conversely π is definable in terms of these operations.*

Idea of the Proof. We construct a model of the geometry and a mapping of the model to itself, which is an automorphism for the operations, but not for the Pieri relation π. If π were definable from the operations, the map would also have to be a π-automorphism. This method of showing the non–definability of a concept goes under the name of "Padoa's method"; Beth has shown that in principle it can always be applied for elementary theories. Thus for non-definability there always exists a suitable model and an automorphism.

Preparation. Let A be a subfield of $\mathbb{C}$, the field of complex numbers. A *Hamel Basis* for $\mathbb{C}$ with coefficients in A is a set $B \subseteq \mathbb{C}$ such that

(i)   each $z \in \mathbb{C}$ can be expressed as a sum $\sum a_i \cdot b_i$, $a_i \in A$, $b_i \in B$, with only finitely many terms, and

(ii)   if $\sum a_i \cdot b_i = 0$, then $a_i = 0$ for all i.

There exists a Hamel basis for each A. To prove this, let $\mathbb{C}$ be well–ordered, $\mathbb{C} = \{c_0, c_1, c_2, \ldots, c_\alpha, \ldots\}$.

c_α will be added to B if c_α cannot be expressed as a finite sum of $c_\beta \in B$ with $\beta < \alpha$. In this construction (with so–called transfinite recursion) one can freely dispose of an initial finite portion of the elements of B, e.g. $b_0 = 1, b_1$ real $\notin A, b_2$ arbitrary complex $\notin A$. We will use this below. B constructed in this way is a basis: Let $c \in \mathbb{C}$. Then $c = c_\alpha$ for some α. If $c_\alpha \in B$ we are done, otherwise by construction of B there exists a representation $c_\alpha = \sum a_i \cdot b_i$. Secondly, let $\sum a_i \cdot b_i = 0$ and $a_i \neq 0$ for one of the a_i's. Let $c_\alpha = \max\{b_j : a_j \neq 0\}$ – this exists because here we have only a finite sum. Then $c_\alpha \in B$, but c_α is representable as a linear combination of earlier basis elements – contradiction. Finally we observe that the representation of elements of $\mathbb{C}$ by linear forms is unique (up to order). Otherwise there exists a contradiction to (ii).

Proof. We assume that we have one or more such operations ρ_i, $i = 1, 2, \ldots, n$, which we write as binary operators between the arguments

$$x \rho_i y = z.$$

These operations must be preserved by the similarity transformations of the Euclidean plane, because this holds for π, and the operations are definable from π. Now consider this plane as the field of complex numbers. Then from the invariance of ρ_i under translation it follows that

$$x\rho_i y = z . \equiv . (x - y)\rho_i 0 = x - y .$$

The unary operations σ_i are introduced as

$$\sigma_i(x) := x\rho_i 0 .$$

Because of the invariance of σ_i under scaling, for all $c \in \mathbb{C}$ we have

$$\sigma_i(x) = y . \equiv . \sigma_i(c \cdot x) = c \cdot y ,$$

hence $\sigma_i(c \cdot x) = c \cdot \sigma_i(x)$ for all $c \in \mathbb{C}$. Therefore if we put $\sigma_i(1) = s_i$, then σ_i reveals itself as multiplication by s_i:

$$\sigma_i(x) = \sigma_i(x \cdot 1) = x \cdot \sigma_i(1) = s_i \cdot x .$$

In other words: for each operation ρ_i there exists some $s_i \in \mathbb{C}$ with

$$x\rho_i y = s_i(x - y) + y .$$

Now let A be the field $\mathbb{Q}(s_1, s_2, \ldots, s_n)$, where $\mathbb{Q}$ is the field of (complex) rational numbers, and let B be a Hamel basis for $\mathbb{C}$ with coefficients in A such that $1, c, d \in B$, c real, d complex. Let T be the linear transformation of $\mathbb{C}$ to itself defined by

$$T(c) = d; \ T(d) = c; \ T(b_i) = b_i \text{ for all other } b_i \in B,$$
$$T(\textstyle\sum a_i b_i) = \sum a_i \cdot T(b_i).$$

T preserves all the operations ρ_i:

$$T(x\rho_i y) = T(s_i \cdot (x - y) + y) = s_i(T(x) - T(y)) + Ty = T(x)\rho_i T(y).$$

By contrast, T does not preserve the relation π. Let us consider the points $0, 1, d$ of the complex plane and the (uniquely determined) circle through $0, 1$ and d with center u.

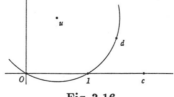

Fig. 2.16.

We observe that $\pi 0u1 \wedge \pi 1ud$. But conversely it is not true that $\pi T(0)T(u)T(1) \wedge \pi T(1)T(u)T(d)$, because by definition of T one would then have $\pi 0T(u)1 \wedge \pi 0T(u)c$. But this last implies that $0, 1, c$ lie on a circle with centre $T(u)$, which contradicts the assumptions. $\square$

The situation is quite different when, instead of points, one considers lines as the basic elements in geometry.

Theorem   *In the Euclidean plane it is possible to define a binary operation on lines, so that using it as unique basic concept, elementary plane Euclidean geometry can be completely axiomatized.*

Sketch of the proof. Intuitively we regard points as pairs of lines positioned perpendicular to each other. The promised binary operation on lines is the reflection:
$$a \cdot b = c$$
means that *c results from the reflection of b in a*.
Perpendicularity expresses itself as
$$a \perp b := . a \cdot b = b \wedge a \neq b.$$

Points are line pairs (a, b) with $a \perp b$; a point (a, b) lies on the line c if
$$(a, b) \in c := a \cdot (b \cdot c) = c,$$

which is apparent from the following figure:

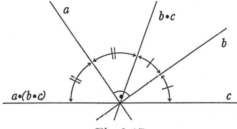

Fig. 2.17.

Two points are equal to each other if
$$(a, b) = (a', b') :\equiv \forall c (a \cdot (b \cdot c) = c . \equiv . a' \cdot (b' \cdot c) = c).$$

Finally the Pieri relation π can be described as follows:
$$\pi(y_1, y_2)(x_1, x_2)(z_1, z_2) :\equiv \quad \exists a \exists b \exists c [(x_1, x_2) \in a \wedge (y_1, y_2) \in a$$
$$\wedge (x_1, x_2) \in b \wedge (z_1, z_2) \in b \wedge (x_1, x_2) \in c$$
$$\wedge c \cdot a = b \wedge (c \cdot y_1, c \cdot y_2) = (z_1, z_2)].$$

This is clear from the following diagram:

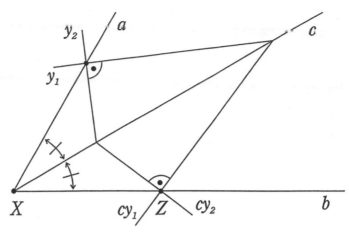

Fig. 2.18.

Thus the operation $a \cdot b = c$ can serve as the basis of elementary plane Euclidean geometry. An explicit complete axiomatization would be obtained by faithful reformulation of the axioms I - V_T in this reduced vocabulary. We have not done this – it just would not be a reasonable approach. Instead, it would be better to total up the naturally appearing requirements for the operation, say also through a coordinatization programme. Then, in fact, one does not find oneself in a new world; indeed there is a methodological link with an alternative foundation of geometry, namely the group theoretic (see for example the book of F. Bachmann).

In elementary geometry, *the limitation of the language* is just as much a phenomenon as in the elementary theory of the real numbers. Thus the reader cannot expect that every geometric concept can be expressed by a formula in the language of elementary geometry. A neat example is the concept of commensurability. One says that two segments AB and CD are commensurable, in symbols $AB \sim CD$, if AB can be subdivided in m subsegments, CD in n, all of which are congruent to each other. The multiples m and n give the length of both segments in terms of the length they have in common, that is the known subsegments. Clearly $AB \sim CD$ is a geometric relation of four points. Yet it is not elementary.

We prove this as follows with the help of non–standard models. Assume that $K(A, B, C, D)$ is an elementary formula, which is equivalent in all models of Γ to $AB \sim CD$. In the standard model we have

$$\forall ABCDE \exists F. CDE \wedge DFE \wedge K(A, B, C, F),$$

that is every extension of the segment CD by a segment DE contains a point F with $CF \sim AB$. But if K were an elementary formula this result would have to hold in the non-standard model of Γ. Here this is not so, for example not when AB is finite and CD infinite.

Of course the language of geometry can be enriched, just as we have enriched the language of the real numbers in Chapter I. We do not want to do this here (see, however, for example, the quoted work of Tarski).

Further Reading

Beth, E. & Tarksi, A.: Equilaterality as the Only Primitive Notion of Euclidean Geometry, Proceedings of the Koninklijke Nederlandse Akademie van Wetenschappen, Series A, vol. 59, pp. 462-467, (1956)

Scott, D.: A Symmetric Primitive Notion for Euclidean Geometry, Proceedings of the Koninklijke Nederlandse Akademie van Wetenschappen, Series A, vol. 59, pp. 456-461, (1956)

Bernays, P.: Die Mannigfaltigkeit der Direktiven für die Gestaltung geometrischer Axiomensysteme, in: Henkin, Suppes & Tarski: The Axiomatic Method, pp. 1-15, Amsterdam, North-Holland, 1959

Tarski, A.: What is Elementary Geometry?, in Henkin, Suppes & Tarski: The Axiomatic Method, pp. 16-29, Amsterdam, North-Holland, 1959

Scott, D.: Dimension in Elementary Euclidean Geometry, in: Henkin, Suppes & Tarski: The Axiomatic Method, pp. 53-67, Amsterdam, North-Holland, 1959

Robinson, R. Binary Relations as Primitive Notions in Elementary Geometry, in: Henkin, Suppes & Tarski: The Axiomatic Method, pp. 68-85, Amsterdam, North-Holland, 1959

Royden, H.L.: Remarks on Primitive Notions for Elementary Euclidean and Non-Euclidean Plane Geometry, in: Henkin, Suppes & Tarski: The Axiomatic Method, pp. 86-96, Amsterdam, North- Holland, 1959

Schwabhäuser, W. & Szczerba, L.W.: Relations on Lines as Primitive Notions for Euclidean Geometry, Fundamenta Mathematicae, vol. 82, pp. 347-355, (1975)

Henkin, L.: Symmetric Euclidean Relations, Proceedings of the Koninklijke Nederlandse Akademie van Wetenschappen, Series A, vol. 65, pp. 549-553, (1962)

Pieri, M.: La geometria elementare istituita sulle nozioni di 'punto' e 'sfera', Memorie di Matematica e di Fisica della Società Italiana delle Scienze, ser. 3, vol. 15, pp. 345-450, (1908)

Bachmann, F. Aufbau der Geometrie aus dem Spiegelungsbegriff, Berlin, Springer-Verlag, 1959

Schwabhäuser, W., Szmielew, W., Tarski, A.: Metamathematische Methoden in der Geometrie, Springer-Verlag, 1983

§ 4 Geometric Constructions

The traditional description of the theory of geometric constructions starts out by listing the different so–called constructional methods: ruler, compass, dividers, parallel ruler, permitted curves and the like, and aims to set down theorems on the possibility of applying these tools and their limitations. The best known are theorems of the impossible kind (trisection of the angle, doubling of the cube, etc.) and theorems about the substitutability or removability of constructional methods (construction with the compass alone). The first apply algebraic methods (Galois theory), the second, clever geometric constructions. The description of the theory of geometric constructions in algebra texts is certainly adequate from the algebraic viewpoint, but the basic concepts must be sharpened for foundational investigations. To this end the circle of ideas from programming languages – which I may here presume – can nowadays give useful service.

The basic constructive operations express themselves as *assignments* in programming languages, for example we shall apply

$$\ell := L(P, Q).$$

This instruction takes two points denoted P and Q, constructs the line joining them, and labels this as ℓ. Assignments are accompanied by *decision operations*, which determine a truth value for given values of the variables, for example whether PQR holds or not. These enter into the "structured" programming language in the contexts

if PQR then ... else ... ;

while $\neg PQR$ do ... .

However in many places, particularly in algebraic descriptions of the theory, decision operations are just forgotten! Presuming a PASCAL–like programming language, it is possible to give a set of constructional methods by listing the variables, constants, assignments and decision operations. This suffices for the definition of the class of entities constructible by means of algorithms. Certainly the simplest such listing is that which describes constructions "with ruler alone".

Affine Constructions ("Ruler Alone")

Variables:

P, Q, R	Points
l, m, n	Lines

Constants:

O, E, E'	Triangle formed by the axes

Assignments:

$Q := P(g, h)$	Intersection point of g and h if $g \nparallel h$
$\ell := L(P, Q)$	Line joining P and Q, if $P \neq Q$
$\ell := L(P, g)$	Parallel to g through P, if $P \notin g$

Decision Operations:

$P = Q, g = h$	Equality
$g \parallel h$	Parallelism
$P \in g$	Incidence

The construction methods allow us, starting from certain initial figures, i.e. the association of lines and points to certain variables, to construct other figures. For example they suffice to realize the geometric additions and multiplications of § 2. The resulting programme (for addition) looks as follows:

```
begin if A = 0  then  G := B  else
     begin  g    :=  L(O, E);
            g'   :=  L(O, E');
            g''  :=  L(E', g);
            h    :=  L(E', A);
            if  B = 0  then  G := A  else
                 begin    j   :=  L(B, g');
                          F   :=  P(j, g'');
                          k   :=  L(F, h);
                          G   :=  P(k, g);
                 end;
     end;
end.
```

From the programmability of the field operations follows: The set of the coordinate values of the constructible points in the Euclidean plane form a field, which includes the field of rational numbers. If just the construction methods above are allowed, this field is exactly $\mathbb{Q}$ (because the coordinates of points of intersection of lines with rational coordinates are again rational etc.). If an initial figure is given, then the field is the one obtained from $\mathbb{Q}$ by adjoining the coordinates of the elements of the initial figure. Therefore:

Theorem 1a *A point P or a line g is constructable from an initial figure using ruler alone if and only if its coordinates belong to the field of rational numbers extended by the coordinates of the initial figure.*

Against this algebraic characterization we want to set a more axiomatic one and ask: what is the connection between the provability of the existence of a point or line and its constructability? Certainly all the points and lines required in the existence statements of Axioms I, D, P and IV^* are constructable; this statement extends - by induction on the length of the proofs - to all theorems provable from these axioms of the form

$$\exists\, PA(P), \;\; \exists\, gB(g),$$

for quantifier free $A(.)$ and $B(.)$. Conversely each construction step involves some existence statement of the form above, provable from I, D, P, IV^*, and we have

Theorem 1b *A formula of the form $\exists\, PA(P)$ or $\exists\, gB(g)$ with quantifier free $A(.), B(.)$ is provable from I, D, P, IV^* if and only if the point or line satisfying the formula is constructable with ruler alone.*

The theorem just proved throws up the related question: which are the construction methods which in this sense determine the axiom system I,II,III,IV*. We find them by checking the existence requirements in the axioms, and find, following Hilbert, constructions with ruler and gauge.

Pythagorean Constructions ("Ruler and Gauge")

In addition to the affine construction methods we have:

$P := E(A, B; C, D)$ 　　　　　　for $A \neq B$ and $C \neq D$ P is the point
　　　　　　　　　　　　　　　　lying on CD with $CDP \wedge AB \approx DP$

PQR 　　　　　　　　　　　　Determination of "lying between".

Theorems 1a and 1b can again be proved; instead of the field of rational numbers in both cases one uses the concept of a *pythagorean ordered field*, that is an ordered field which along with　a　and　b　always contains $\sqrt{a^2 + b^2}$. Essentially the proofs are contained in § 36 and § 37 of Hilbert's little book.

Theorem 2a *A point or a line is constructable from an inital figure with ruler and gauge if and only if its coordinates belong to the extension of the ordered pythoagorean field over the rational numbers obtained by adjoining the coordinates of the initial figure.*

Theorem 2b *A formula of the tpye* $\exists P A(P)$ *and* $\exists g B(g)$ *with quantifier free* $A(.) B(.)$ *is provable from I,II,III,IV* if and only if the points (or lines) in the formula are constructable with ruler and gauge.*

The classical or "platonic" construction methods, ruler and compass, involve a somewhat different extension of the affine construction methods.

Euclidean Constructions ("Ruler and Compass")

In addition to the affine construction methods we have:

$P \in \langle Z; A, B\rangle$ $\qquad\qquad$ P lies in the interior of the circle about Z with radius $\overline{AB}$, for $A \neq B$

$P_1, P_2 := P(Z; A, B; C, D)$ $\qquad$ In the case that $C \in \langle Z; A, B\rangle$ and $C \neq D, P_1$ and P_2 are the two intersection points of the line CD with the circle $\langle Z; A, B\rangle$

In order to formulate the corresponding Theorems 3a and 3b we need two concepts. First an Euclidean ordered field is an ordered field which for each positive element a also contains the square root $\sqrt{a}$. It is clearly possible to construct the square root with ruler and compass; one takes the known construction for division according to the proportion

$$a : x = x : 1,$$

which follows from known relations for the right–angled triangle. Thus:

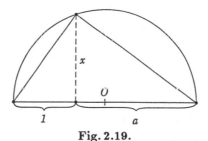

Fig. 2.19.

Secondly, the *circle intersection axiom* K, which says that a line through a point lying inside a circle has at least one intersection point with the circle (in fact it has two).

Theorem 3a *A point or a line is constructable from an initial figure with ruler and compass if and only if its coordinates belong to the extension of the ordered Euclidean field over the rational numbers obtained by adjoining the coordinates of the initial figure.*

Theorem 3b *A formula of the type* $\exists P A(P)$ *or* $\exists g B(g)$ *with quantifier free* $A(.), B(.)$ *is provable from* I,II,III,IV* *and* K *if and only if the points (or lines) in the formula are constructable with ruler and compass.*

Given what has been said already we can spare ourselves the details of the proofs. It is interesting to note that from the completeness axiom V (i.e. V_T) we only need the special case expressed in the circle intersection axiom K.

That K is indeed an additional requirement can easily be seen from the existence of non–Pythagorean Euclidean fields; $\sqrt{2 \cdot \sqrt{2} - 2}$ is not contained in the Pythagorean field over $\mathbb{Q}$.

The completeness axiom V, and in particular the Archimedean axiom which it implies, is of particular importance in the theory of geometric constructions. For example, one might try to equate Kürschak's proposed constructions with ruler and unit of length (see Hilbert § 36) with Pythagorean constructions.

Ruler and Unit of Length

Variables
P, Q, R Points
Constants
O, E, E' Triangle formed by the axes

Decision Operations
$A = B$:   equality
$ABCD$:   holds if and only if, either $A \neq B$ and $C \neq D$ and the intersection point P of the lines AB and CD is either A, B, C or D, or P is such that APB or CPD, or alternatively $A \neq B$, $C = D$ and ACB.

Assignments
$P := U(A, B)$:   For $A \neq B$ P is the point lying on AB with
                       $ABP \wedge BP \approx OE$.
$P := I(A, B; C, D)$:   In the case that $ABCD$ and $C \neq D$, P is
                       the intersection point of AB and CD.

These tools can probably not be further reduced in any essential way, but serve – as one convinces oneself – for the construction in all problems solvable by means of ruler and gauge. We exhibit this fact on a particular construction – the construction of the intersection point of the lines given by the pairs A, B and C, D.

```
begin
    while ¬ABCD do
        begin  D := U(C, D); C := U(D, C)  end;
    P := I(A, B; C, D)
end .
```

We have spelt out this method of construction in such detail, because it does not yet quite give what has just been promised. Although all elementary exercises of construction with ruler and gauge can be carried out, we have made an implicit assumption concerning the scope of the axioms I–IV*, even the scope of the axioms I–V_T. Namely: The programme for the general determination of intersection points does not need to deliver an intersection point in all models of I V_T! One has only to consider a *non–standard model* of I–V_T with infinitesimal elements. In such a model there always exists a pair of intersecting lines which form an infinitesimal angle. Then the Archimedean principle of repeated use of the unit of length applied in the programme will never end in generating the situation $ABCD$; although it exists, the intersection point is not constructable.

Finally we consider methods of construction which involve the geometry described in § 3, the one based solely on the concept of a line and a binary operation.

Pythagorean Constructions   ("Paper folding")

Variables:

$l, m, n, g, h \ldots$ lines

Constants:

x, y, u tho lines OE, OE' and FF'

Assignments:

$\ell := \bot(g, h):$ for $g \nparallel h$ the perpendicular to g through the point of intersection

$\ell_1, \ell_2 := L(g, h):$ if $g \nparallel h$ the angle bisectors, if $g \parallel h$ the mean parallel line

$\ell := L(g_1, g_2; h_1, h_2):$ if $g_1 \nparallel g_2$ and $h_1 \nparallel h_2$ the line joining the points of intersection if distinct

Decision Operations:

$g = h:$ equality

$(a_1, a_2)(b_1, b_2)(c_1, c_2):$ a_1 and a_2 meet in A, b_1 and b_2 in B, c_1 and c_2 in C, and ABC holds.

One thinks of lines as being given as folds in a piece of paper, and ascertains that the basic constructions can really be carried out through such paper folding operations.

Theorem   *By folding paper one can carry out exactly the same constructions as are possible with ruler and gauge.*

Proof. We must show that the basic constructions of each construction gemetry can be replaced by the other.

(a) *Simulation of Ruler and Gauge Constructions*

1) $\ell := L(P, g)$, where P is the intersection point of p_1, p_2 and $P \notin g$.
We use the harmonic position of the points $A, B, C, \text{"}\infty\text{"}$.

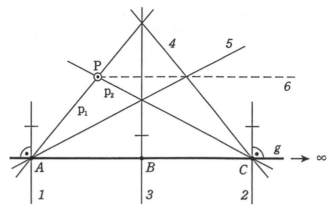

Fig. 2.20.

The sequence of paper folds in the construction is indicated by the numbering of the lines.

2) $P := E(A, B; C, D)$

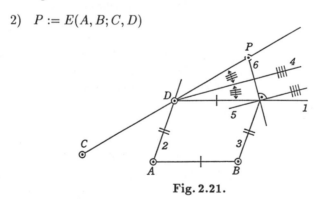

Fig. 2.21.

3) $P \in h$, where P is the intersection of p_1, p_2, if and only if $\bot(h, p_1) = \bot(h, p_2)$.

4) $P = Q$, where P is the intersection of p_1, p_2 and Q the intersection of q_1, q_2 if and only if $P \in q_1$ and $P \in q_2$.

5) $g \parallel h$ holds for $g \neq h$ if and only if $\bot(h, x) = \bot(g, \bot(h, x))$.

6) If A, B, C are respectively the intersection points of $a_1, a_2; b_1, b_2; c_1, c_2$, then ABC holds if and only if $E(A, B; C, B) = A \wedge E(C, B; A, B) = C$, where E has already been constructed above.

(b) *Simulation of Paper Folding Constructions*

1) x, y, u are generated with $L(O, E), L(O, E'), L(E, E')$.

2) $\ell := \perp(g, h)$. Here we apply the theorem about the orthocenter.

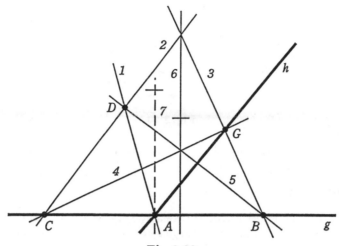

Fig. 2.22.

The line 1 through A will be so positioned through O, E or E' that it is distinct from g and h. The rest of the construction is indicated by the numbering. The segments AB, AC, AD and AG are congruent to OE.

3) $\ell := L(g, h)$.
First case: $g \nparallel h$:

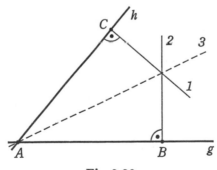

Fig. 2.23.

The segments AB and AC are congruent to OE.

Second case: $g \parallel h$:

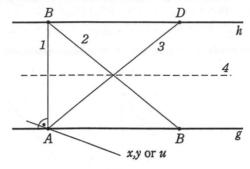

Fig. 2.24.

The segments AB, BD and AC are congruent to each other.

Further Reading

Engeler, E.: Remarks on the Theory of Geometrical Constructions, in: C. Karp: The Syntax and Semantics of Infinitary Languages, pp. 64-76, Lecture Notes in Mathematics, vol. 72, Berlin, Springer–Verlag, 1967

Engeler, E.: On the Solvability of Algorithmic Problems, Rose & Sheperdson: Logic Colloquium '73, pp. 231-251, Amsterdam, North-Holland, 1975

Seeland, H.: Algorithmische Theorien und konstruktive Geometrie, Stuttgart, Hochschul–Verlag, 1978

Schreiber, P.: Grundlagen der konstruktiven Geometrie, VEB Deutscher Verlag der Wissenschaften, 1984

Chapter III. Algorithmics

In the last chapter it is customary for an author to indulge his hobby a little. This I will also do, but since it is only Chapter III in a thin book, it needs a little more justification.

We have already discussed algorithms in different contexts; once in the introduction when we contrasted classical and constructive mathematics, then in connection with decision procedures in the elementary theory of the real numbers, and finally just above in the form of geometric constructions. The possibilities and limits of effective procedures, i.e. algorithms, in mathematics form just as fundamental a complex of questions as the question concerning the possibilities and limits of the axiomatic description of fundamental mathematical structures. Again our approach is metatheoretic. We will try to speak of algorithms as objects in the same way as previously of real numbers or of geometric objects such as points and lines. Again we introduce basic relations between such objects as elements of speech. Finally languages built from these elements serve us as the means of expression for axiomatic investigations, analogous to those in previous chapters, and also as an alternative formulation of the theory of computation as a theory of formal operation on algorithms.

§ 1   What is an Algorithm?

At the start of the study of algorithms stands the concept of a function, more precisely, the concept of calculable functions.

Of course, as is known, one can base the concept of a function on the *concept of a set*: a function from A to B is a subset $F \subseteq A \times B$ with the property that

$$\forall x \in A \, \exists y! \in B . \langle x, y \rangle \in F .$$

According to this relational definition the function is given as a whole, spread out in full *extension* before us. However this idealized set theoretic description is not really appropriate as a practical concept of a function, which places calculation at the forefront. A function is a *computation rule*, which, applied to an argument, delivers a well–defined value. In schoolwork these rules mostly take the form of a formula, more precisely, of a functional expression such as

$$(x + 2) \cdot (x - 2)$$

or

$$(x^2 - 4).$$

These two functional expressions f_1 and f_2 give the same value on $\mathbb{Q}$; f_1 applied to $x \in \mathbb{Q}$ is always the same as f_2 applied to x. If we write $f * x$ for "f applied to x", then $f_1 \neq f_2$, even though $f_1 * x = f_2 * x$ for arbitrary x. The two expressions differ not only externally, but also in important aspects, for example in the complexity: f_1 requires more calculation steps than f_2. With algorithms we are less interested in *extension*, the range of values of the function, than in *intension*, the inner character. The sense of a functional expression in the theory of algorithms lies in its nature as a recipe for carrying out (a sequence of) operations. Our aim is to develop the theory of computation, or "Algorithmic" as the theory of operations *with* and *on* algorithms. The axiomatic foundation of algorithmic chosen for this purpose is best compared with the axiomatization of group theory: - In mathematics we meet again and again, and in very different contexts, collections of things which form a group. In the same way we come across many closed collections of operational prescriptions, for example geometric constructions, various programming languages, recipes for formal operation with algebraic or logical expressions etc. Where group theory works with the concept of composing group elements, the fundamental operation in axiomatic algorithmic is that of *application* $f * a$. Above all the theory ought to display what is common to such collections of algorithms; this is brought closer to the reader in the following considerations.

In the beginning is the concept of a function, understood as a computational rule. Clearly, rules or algorithms can also arise as arguments for rules or be changed by such into others. And now we make the great loop into the unknown: We consider rules and algorithms as being collected together in one undifferentiated set. This point of view differs from that, say, of analysis, where we distinguish between functions from $\mathbb{R}$ to $\mathbb{R}$, functionals from $\mathbb{R}^\mathbb{R}$ to $\mathbb{R}$ and functionals of higher type, say, from $\mathbb{R}^\mathbb{R}$ to $\mathbb{R}^\mathbb{R}$ etc. The proposed disregarding of types is related to the viewpoint of set theory, which similarly makes no distinctions between the levels of sets.

We recall that in set theory the ordered pair of two sets x and y is defined to be the set $\{\{x\}, \{x, y\}\}$, and hence is an element of the power set of the power set of the particular set, say $\mathbb{R}$, from which x and y come. By means of the extensional concept of a function, each function from $\mathbb{R}$ to $\mathbb{R}$ becomes a set, viz. an element of the power set of the set of pairs defined above. And real numbers are simply certain sets, as are functions and functionals. In the manner just sketched using set theory with the one basic notion "x is an element of y", one is able to build up the structurally extremely complicated totality of mathematical concepts. This methodological paradigm is now to be made useful for the range of objects which we here have in view.

The disregarding of types in the collection of algorithms again allows economy with the basic concepts of the axiomatization, just as it does in axiomatic set theory. Thus we do not need to distinguish between rules and algorithms on one and several variables. We can express a rule $F(x, y)$ with two parameters as a rule f, which applied to x gives $f * x$, which in turn applied to y gives $F(x, y)$. This idea goes back to Frege, and was first applied systematically by Schönfinkel; he illustrates the idea with the help of the numerical function $x - y$ als follows. Consider the expression as a recipe for y alone, then this has the form "take the difference between x and what follows", which we denote by $(x-\ )$. The rule f applied to x therefore yields this $(x-\ )$. Hence:

$$(f * x) * y = (x-\ ) * y = x - y.$$

Clearly this idea carries over to many–variable functions – thus $F(x_1, x_2, \ldots, x_n)$ can be expressed as $((\ldots (f * x_1) * x_2) \ldots * x_n)$ for an appropriate f.

Convention. In order to simplify the notation we will often omit the symbol "$*$", and economize in the use of brackets, thus:

$$(fxyz) \text{ is } (((f * x) * y) * z).$$

The instruction "first carry out rule f and then apply g to the result" is itself a rule, the *composition*, which makes f and g into the rule usually denoted by $(f \circ g)$. The latter operates on x as follows:

$$(f \circ g) * x = f * (g * x).$$

This way of combining f and g is independent of the special properties of the functions and values involved; it is itself a rule, and as such ought to belong to our range of objects. Put another way, we must have an object $\boldsymbol{B}$, which satisfies the relation

$$\boldsymbol{B} fgx - f(gx),$$

and this for arbitrary f, g and x.

The expression

$$\sin(a + \tan(b)) + \tan(\sin(a + b))$$

represents a rule or algorithm $F(\sin, \tan, +, a, b)$, which combines the functions $\sin, \tan$ and $+$, as well as the numerical values a, b in a certain way. The form of the combination is again independent of the special properties of the functions and values occurring; as an algorithm it ought itself to belong to the range of objects. Let us make this requirement clear. The expression above uses two unary function symbols $\sin$, $\tan$ and one binary function symbol $+$, which is, as explained already, to be replaced by the unary $\oplus$:

$$\oplus xy \text{ is } x + y.$$

The expression above, now reformulated on the basis of the operation of application, is to be written as

$$\oplus(\sin(\oplus a(\tan b)))(\tan(\sin(\oplus ab)))$$

and we require the existence of an object f with

$$f \sin \tan \oplus ab = \oplus(\sin(\oplus a(\tan b)))(\tan(\sin(\oplus ab))).$$

Moreover, for arbitrary objects x, y, z, u, v

$$fxyzuv = z(x(zu(yv)))(y(x(zuv))).$$

In this way we turn the form and manner of *combining* sin, tan, etc. *as such* into a mathematical entity, without bringing properties of these objects into play. We say that "f results from functional abstraction".

What is common to these examples is the following: we consider certain objects, "rules" or "algorithms", as given. These are combined with each other solely with the help of the operation of application,

$$f * (g * x) \quad \text{or} \quad z * (x * (z * u) * (y * v)) * (y * (x * ((z * u) * v))).$$

And then we require the existence of an object such as **B** above or f here which realizes the combination of algorithms itself as an algorithm by means of functional abstraction.

We now set out the possibility of functional abstraction in general as an axiomatic requirement for our range of objects.

Combinatory Algebras

Let $\mathcal{D} = \langle D, *, c_1, \ldots, c_m \rangle$ be an algebraic structure with non–empty basic set D, having distinguished elements $c_1, c_2, \ldots, c_m, m \geq 0$, and a binary operation $*$. Let $t(x_1, \ldots, x_n)$ be a term in D, i.e. t is built up from constant symbols for the distinguished elements, the variables $x_1, \ldots, x_n$ by means of brackets and the operation symbol $*$. An element f from D *represents* t in $\mathcal{D}$, if for each choice of elements $a_1, \ldots, a_n$ in D, we have

$$fa_1, \ldots a_n = t(a_1, a_2, \ldots, a_n).$$

The algebra $\mathcal{D}$ is *combinatorially complete*, if each term t can be represented in $\mathcal{D}$. Then $\mathcal{D}$ is called a combinatory algebra, which is non–trivial, if D contains more than one element.

The combination given by a term t is therefore always represented in a combinatory algebra by (at least) one object. Given their origin we call such objects *Combinators*.

Schönfinkel (1924) and independently Curry(1930) posed the mathematically stimulating question whether or not all combinators could be put together from just a few. This is indeed so:

S and **K** suffice

Let $\mathcal{D} = \langle D, *, c_1, \ldots, c_m \rangle$ be an algebraic structure, in which there are two elements **S** and **K** , which identically satisfy the following equations, i.e. for all $x, y, z \in A$:

$(S)\ \boldsymbol{S}xyz = xz(yz)\,,$

$(K)\ \boldsymbol{K}xy = x\,.$

Then $\mathcal{D}$ is a combinatory algebra. (Conversely it is clear that in each combinatory algebra there are always elements **S** and **K** which satisfy the equations above.)

The proof of the theorem above consists of a simple induction. We carry it out here, because it shows that the origin of the combinators **S** and **K** is not so enigmatic.

Lemma   *For each n and each term $t_n(x_1, \ldots, x_n)$ in **S**, **K** and $x_1 \ldots, x_n$ there exists a term $t_{n-1}(x_1, \ldots, x_{n-1})$ in **S**, **K** and $x_1, \ldots, x_{n-1}$ so that*

$$t_{n-1}(x_1, \ldots, x_n) * x_n = t_n(x_1, \ldots, x_n)\,.$$

Proof. If the case that t_n only consists of **S** and **K** or x_i with $i \neq n$, we take $\boldsymbol{K}t_n$ for t_{n-1}; then $t_{n-1}x_n = (\boldsymbol{K}t_n)x_n = t_n$. In the case that t_n is composite, say $t_n = l'_n l''_n$, we apply the inductive assumption. Thus suppose $t'_n = t'_{n-1}x_n$ and $t''_n = t''_{n-1}x_n$. Take $\boldsymbol{S}t'_{n-1}t''_{n-1}$ for t_{n-1}, and check

$$t_{n-1}x_n = \boldsymbol{S}t'_{n-1}t''_{n-1}x_n = t'_{n-1}x_n(t''_{n-1}x_n) = t'_n t''_n = t_n\,.$$

The case $t = x_n$ remains, for which we need a combinator **I** with

$$\boldsymbol{I}x = x\,.$$

One can put together such a combinator from **S** and **K** in various ways, for example as $\boldsymbol{SKK}$ (Boskowitz) or $\boldsymbol{SK}(\boldsymbol{KK})$ (Bernays), as is easily verified. □

Finally combinatory completeness follows by repeated application of the lemma:

$$t_n(x_1,\ldots,x_n) \;=\; t_{n-1}(x_1,\ldots,x_{n-1})x_n = t_{n-2}(x_1,\ldots,x_{n-2})x_{n-1}x_n =$$
$$\ldots \;=\; t_0 x_1 x_2 \ldots x_n \,.$$

As an example it may amuse the reader to put together the combinators introduced above, using S and K. Thus

$$B = S(KS)S\,.$$

In the same way as for I there are many variants; it is also amusing to look for algorithms which make this exercise mechanical.

Further Reading

Curry, H.B.: Grundlagen der kombinatorischen Logik, Am. J. of Math. 52,
    pp. 509-536, 789-834, (1930); see also the books with coauthors: Combinatory Logic,
    vols. I, II, Amsterdam, North-Holland, 1958, 1972
Frege, G.: Begriffschrift, Halle a.S., 1879, (cf. in particular § 9). 2. edition: Hildesheim,
    G. Olms Verlag, 1977
Schönfinkel, M.: Über die Bausteine der mathematischen Logik, Math. Annalen 92,
    pp. 305-316, (1924); in an addendum (by H. Behmann) there is an error, later
    realized by its author.
Quine, W.v.O.: On the building blocks of mathematical logic (= translation and
    commentary of the above paper), in: van Heijenoort, ed.: From Frege to Gödel.
    Cambridge, Mass., Harvard University Press, 1967

§ 2 The Existence of Combinatory Algebras: Combinatory Logic

If the reader has tried to manufacture a useful example of a combinatory algebra for himself, at some stage he will blame me for doing away, along with type differences with a great deal of mathematical intuition. Indeed in this connection the step from the possible to the contradictory has been made by various mathematicians. From the point of view of the founders (Schönfinkel, Curry and Church) the general aim was not just the axiomatization of the concept of application for general functions, but rather a functional foundation of all logic and mathematics. In particular Curry and Church originally proposed systems, which together with reasonable appearing ingredients for combinatory algebra, also included logical rules and parts of mathematics. These expanded systems then proved to be inconsistent (Kleene & Rosser 1935). Curry considered that this lay in the nature of the subject – he compared the axioms with Frege's system and gave the opinion that "here we are dealing with notions of such great generality, that intuition fails us. We are researching in a no–man's land between what is certain and what is known to be contradictory".

The central question as to the existence of non–trivial combinatory algebras can be approached from different points of view. Historically the first of these is the *proof theoretic*. It was followed both by Church and Curry under the influence and along the lines of Hilbert's programme (for proving number theory and analysis consistent). We want to describe this approach, and then turn to the second, the *algebraic*.

The proof–theoretic approach considers the mathematical theory of combinatory algebras as a formal–deductive discipline. As such it starts out from a system of axioms and deductive rules, and the question of the existence of such algebras is replaced by the question of the freeness from contradiction of the formal system – combinatory algebra becomes combinatory logic.

Combinatory Logic

Intended Structure:
$\mathcal{D} = \langle D, *, \; S, K \rangle$

Language:
Terms, built up from *atomic terms*, i.e. variables $x, y, z \ldots$ and constants S, K with the help of the binary operation $*$.
Formulae: equations between terms.

Axioms:
$t = t$ for atomic terms,
$St_1t_2t_3 = t_1t_3(t_2t_3)$, $Kt_1t_2 = t_1$ for arbitrary terms t_1, t_2, t_3.

Deductive Rules:

$$\frac{t_1 = t_2}{t_1t_3 = t_2t_3} \qquad \frac{t_1 = t_2}{t_3t_1 = t_3t_2} \qquad \frac{t_1 = t_2}{t_2 = t_1}$$

$$\frac{t_1 = t_2 \qquad t_2 = t_3}{t_1 = t_3}$$

Provability
A sequence $\alpha_1, \alpha_2, \alpha_3 \ldots \alpha_n$ of equations is called a proof, when for each
$i = 1, 2, \ldots, n$ either α_i is an axiom, or there exists some $j, k < i$,
so that α_i follows from α_j and α_k by means of one of the deductive rules. The equation α_n is then called provable.

Since clearly the language of combinatory logic is drastically limited, in it there does not exist the usual concept of logical contradiction, nor then the deducibility of both a statement A and its negation $\neg A$. Observe that in the usual logic from A and $\neg A$ one can deduce any formula. We wish to promote this fact to a definition and more generally to say, that a logical system is contradictory, if all formulae in the language can be deduced. Combinatory logic is then shown to be contradiction–free, if we can show that there exists no sequence $\alpha_1, \alpha_2, \ldots, \alpha_n$, which constitutes a proof, and where let us say α_n is the equation $\boldsymbol{K} = \boldsymbol{S}$.

The important thing in the proof–theoretic approach is just that it frees us from the content of the existence question. We pose a purely formal question, i.e. one that concerns exclusively a proof structure built up from a finite collection of formulae: does there exist a sequence of α's with such and such properties? And in order to answer this question it suffices, in this case, to use simple methods of proof and statements about formal manipulations on the proof structure, see in particular Lemma 4 below. We will present these manipulations with the help of a reduction calculus, which is of independent interest (as will be indicated in §5).

The reduction calculus starts from the remark, that the axioms for $\boldsymbol{S}$ and $\boldsymbol{K}$ possess a natural asymmetry: the left side can be regarded as the argument, the right side as the result of a reduction. The repeated application of such reductions is here presented as derivation inside a system of axioms and deductive rules, which arise quite naturally.

Reduction Calculus

Language:
Terms as in combinatory logic,
Formulae: reducibility statements $t_1 \geq t_2$.

Axioms:
$t \geq t$ for atomic terms,
$\boldsymbol{S}t_1t_2t_3 \geq t_1t_3(t_2t_3)$, $\boldsymbol{K}t_1t_2 \geq t_1$ for arbitrary terms.

Deductive Rules:

$$\frac{t_1 \geq t_2}{t_1t_3 \geq t_2t_3} \qquad \frac{t_1 \geq t_2}{t_3t_1 \geq t_3t_2}$$

$$\frac{t_1 \geq t_2 \quad t_2 \geq t_3}{t_1 \geq t_3}$$

Provability:
Analogous to combinatory logic.

It so happens that reductions only take place on terms which begin with either $\boldsymbol{S}$ or $\boldsymbol{K}$. Somewhat more general is the notion of contraction: if t contains a subterm of the form $\boldsymbol{S}t_1t_2t_3$ (respectively $\boldsymbol{K}t_1t_2$) and t' results from t by replacing this subterm by $t_1t_3(t_2t_3)$ (respectively by t_1), we say that t contracts to t'. The concept of *simultaneous contraction* is slightly more general, and more suitable for our technical purposes. We write $t | \!\!\mapsto t'$ if in t there exist one or more non-overlapping subterms of the specific form $\boldsymbol{S}t_1t_2t_3$ or $\boldsymbol{K}t_1t_2$,

such that t' results from their simultaneous replacement by $t_1 t_3 (t_2 t_3)$ or t_1.
From this concept one can easily formulate a calculus:

Contraction Calculus

Language:
Terms for combinatory logic
Formulae: $t \| \mapsto t'$.

Axioms:
$t \| \mapsto t$ for atomic t,
$S t_1 t_2 t_3 \| \mapsto t_1 t_3 (t_2 t_3), K t_1 t_2 \| \mapsto t_1$ for arbitrary t_i.

Deductive Rules:

$$\frac{t_1 \| \mapsto t_1', t_2 \| \mapsto t_2'}{t_1 t_2 \| \mapsto t_1' t_2'}$$

Provability:
Again analogously to the previous cases.

Clearly $t \| \mapsto t'$ is provable in the contraction calculus precisely if t' results
from t by *one* simultaneous contraction. In particular the "contraction" $t \| \mapsto t$
is provable for each t.

We now proceed to present the connections between the provability of formulae in combinatory logic, the reduction and the contraction calculi. We begin
with the remark that each reduction can be understood as a sequence of simultaneous contractions. More precisely:

Lemma 1 $t \geq t'$ *is provable in the reduction calculus if and only if one can
find a sequence* $t_0, t_1, \ldots, t_n$ *so that* t_0 *is identical with* t, t_n *with* t', *and each*
$t_i \| \mapsto t_{i+1}$ *is provable in the contraction calculus.*

Proof. If $t \geq t'$ is provable, it is either an axiom, in which case so is $t \| \mapsto$
t', or it is the result of a deductive rule. If we assume the assertion for the
premisses of the deductive rule, then it follows for the conclusion. For example
let $s_1 \| \mapsto s_2 \| \mapsto s_3 \| \mapsto \ldots \| \mapsto s_n$ be the sequence which corresponds to the
premiss $t_1 \geq t_2$ of the first deductive rule of the reduction calculus. Then the
sequence $s_1 t_3 \| \mapsto s_2 t_3 \| \mapsto s_3 t_3 \| \mapsto \ldots \| \mapsto s_n t_3$ is the desired sequence for the
conclusion $t_1 t_3 \geq t_2 t_3$, and similarly for the other rules of deduction.

Suppose conversely that $t_1 \| \mapsto t_2 \| \mapsto \ldots \| \mapsto t_n$ are all provable in the
contraction calculus. We must find a proof of $t_1 \geq t_n$, where because of the
transitivity of $\geq$ it suffices to find a proof for each $t_i \geq t_{i+1}$. Suppose therefore
that $t \| \mapsto t'$ is provable. If it is an axiom we are again finished. If $t \| \mapsto t'$
follows from the contraction rules, say as $t_1 t_2 \| \mapsto t_1' t_2'$, then we once more

argue inductively as follows: by assumption the provability of the premises $t_1 \Vdash t_1'$ and $t_2 \Vdash t_2'$ implies that of $t_1 \geq t_1'$ and $t_2 \geq t_2'$. We therefore infer the provability of $t_1 t_2 \geq t_1' t_2$ and $t_1' t_2 \geq t_1' t_2'$, and so by transitivity that of $t_1 t_2 \geq t_1' t_2'$. $\square$

Lemma 2   *If $a \Vdash m$ and $a \Vdash n$ are provable in the contraction calculus, then from the proof one can effectively find some z, so that $m \Vdash z$ and $n \Vdash z$ are provable.*

Proof. First we take the case that a is an atomic term. Then $a \Vdash m$ and $a \Vdash n$ are axioms, and a, m, n and z are identical.

Suppose now that a is composite, say $a'a''$. Now apply induction on the length of the proof.

If both formulae $a \Vdash m$ and $a \Vdash n$ are axioms, then both begin with S or both begin with K. Then m and n are again both identical, and we can take m as z.

Suppose now that one of the formulæ is an axiom, let us say $a \Vdash m$ with a of the form $Suvw$ (the case Kuv is handled in a similar way) and m is then $uw(vw)$. The other formula results from using the contraction rules from other provable formulæ; in the present case it can only have the form $Suvw \Vdash n'w'$, and the final part of the proof has the form below. (Here we give the structure of the proof, which by definition must simply consist of a sequence of formulæ, presented in the obvious tree–like form.)

$$\frac{\displaystyle \frac{\displaystyle \frac{S \Vdash S \qquad u \Vdash u'}{Su \Vdash \ell \qquad v \Vdash v'}}{Suv \Vdash n' \qquad w \Vdash w'}}{Suvw \Vdash n'w'}$$

Therefore ℓ equals Su' and n' equals $Su'v'$, and we have a contraction proof of $Suvw \Vdash Su'v'w'$. So the term n has the form $Su'v'w'$. Now $Suvw \Vdash uw(vw)$ and $Su'v'w' \Vdash u'w'(v'w')$ are both axioms, and $uw(vw) \Vdash u'w'(v'w')$ follows from $u \Vdash u', v \Vdash v'$ and $w \Vdash w'$. In this way we obtain the diagram

$$
\begin{array}{ccc}
Suvw & \Vdash & Su'v'w' \\
\mathord{\overline{\underline{\top}}} & & \mathord{\overline{\underline{\top}}} \\
uw(vw) & \Vdash & u'w'(v'w')
\end{array}
\qquad i.e. \qquad
\begin{array}{ccc}
a & \Vdash & n \\
\mathord{\overline{\underline{\top}}} & & \mathord{\overline{\underline{\top}}} \\
m & \Vdash & z \,.
\end{array}
$$

Finally suppose that $a \Vdash m$ and $a \Vdash n$ are both proved by simultaneous contraction, say with

$$\frac{a' \Vdash m',\ a'' \Vdash m''}{a'a'' \Vdash m'm''} \qquad \text{and} \qquad \frac{a' \Vdash n',\ a'' \Vdash n''}{a'a'' \Vdash n'n''}.$$

We apply the assumption separately to a' and a''. There then exist z' and z'' with $m' \Vdash z'$, $n' \Vdash z'$ and $m'' \Vdash z''$, $n'' \Vdash z''$. By means of contraction we have $m'm'' \Vdash z'z''$ and $n'n'' \Vdash z'z''$, so that z can be taken to be $z'z''$.   □

Lemma 2 can be immediately carried over to the reduction calculus.

Lemma 3 *If $a \geq m$ and $a \geq n$ are provable in the reduction calculus, then we can find some z with $m \geq z$ and $n \geq z$.*

Proof. By Lemma 1 we have the upper row and the left side of the following diagram:

$$
\begin{array}{ccccccccccc}
a & \Vdash & a_1 & \Vdash & a_2 & \Vdash & \cdots & & \Vdash & a_j & \Vdash & m \\
\bar{\top} & & \bar{\top} & & \bar{\top} & & & & \bar{\top} & & \bar{\top} & \\
b_1 & \Vdash & c_{11} & \Vdash & c_{12} & & & & c_{1j} & \Vdash & e_1 & \\
\bar{\top} & & \bar{\top} & & \bar{\top} & & & & & & \bar{\top} & \\
b_2 & \Vdash & c_{21} & \Vdash & c_{22} & & & & & & & \\
\bar{\top} & & & & & & & & & & & \\
\vdots & & \vdots & & & & & & & & \vdots & \\
b_i & \Vdash & c_{i1} & & & & & & c_{ij} & \Vdash & e_i & \\
\bar{\top} & & \bar{\top} & & & & & & \bar{\top} & & \bar{\top} & \\
n & \Vdash & d_1 & \Vdash & \cdots & & & & \Vdash & d_j & \Vdash & z\,.
\end{array}
$$

Both sides of the diagram are enlarged by repeated application of Lemma 2 as given, and the lemma follows by further application of Lemma 1 to the right side and lower part of the diagram.   □

Lemma 4 *From each proof of the equation $m = n$ in combinatory logic, one can extract some term z effectively for which $m \geq z$ and $n \geq z$ are provable in the reduction calculus.*

Proof. If $m = n$ is an axiom of combinatory logic, then either m and likewise n are atomic terms, and we can take z to be m, or we have one of the two "genuine" axioms, say $Sabc = ac(bc)$. However then both of $Sabc \geq ac(bc)$ and $ac(bc) \geq ac(bc)$ are provable, the latter through repeated application of the first two deductive rules, starting from the axioms $u \geq u$ for atomic u.

In the case when $m = n$ is itself the result of applying a deductive rule, we consider the resulting subcases. Suppose therefore that m is of the form ab, n of the form ac and the deduction is

$$\frac{b = c}{ab = ac}.$$

By the inductive assumption for the premiss there exists some z' with $b \geq z'$ and $c \geq z'$. In the reduction calculus we then obtain $ab \geq az'$ and $ac \geq az'$; with az' we have found a suitable z. The considerations for the second deductive rule are analogous. The third deductive rule is of the form

$$\frac{m = a, a = n}{m = n}.$$

By the inductive assumption there exist u and v with $m \geq u$, $a \geq u$, $a \geq v$, $n \geq v$. Now apply Lemma 3 to $a \geq u$ and $a \geq v$ – there exists some z with $u \geq z$ and $v \geq z$. It follows finally from the transitivity of $\geq$ that $m \geq z$ and $n \geq z$. □

Theorem $S = K$ *is not provable in combinatory logic, i.e. the logic is consistent.*

Proof. By Lemma 4 a proof of $S = K$ would give a z for which $S \geq z$ and $K \geq z$ would be provable. By Lemma 1 it would then be possible to pass from S to K by means of a sequence of contractions to z, which is clearly not possible.

(The proof given above is due essentially to E. Zachos.)

Further Reading

Church, A. & Rosser, J.B.: Some properties of conversion, Transactions Amer. Math. Soc. 39, pp. 472-482, (1936)

Curry, H.B.: Recent advances in combinatory logic, Bull. Soc. Math. Belgique 20 pp. 288-298, (1968); (cf. pp. 296-297).

Engeler, E.: Zum logischen Werk von Paul Bernays, Dialectica 32, pp. 191-200, (1978)

Kleene, pp.C. & Rosser, J.B.: The inconsistency of certain formal logics, Annals of Math. 36, pp. 630-636, (1935)

Zachos, E.: Kombinatorische Logik und S–Terme, Dissertation ETH Zürich, 1978

§ 3 Concrete Combinatory Algebras

First of all I must apologize for the word "concrete"; the concreteness of the model we describe is somewhat comparable with the concreteness of the field of real numbers, as constructed by Dedekind. It is thus concreteness relative to an unreflectively borrowed substratum from naive set–theory – as concrete therefore as the objects of classical mathematics.

If we accept this viewpoint, however, we have in the previous section already created the basis for the construction of a combinatory algebra, i.e. for making concrete a model of combinatory logic. Namely: the terms of combinatory logic are divided by provability into equivalence classes: t and t' are equivalent if the equation $t = t'$ can be proved. In relation to the composition $*$ this equivalence is even a congruence. If $t_1 = t_1'$ and $t_2 = t_2'$ are provable, then so is $t_1 * t_2 = t_1' * t_2'$, as is immediately apparent. So we arrive at the term–model.

This model is not trivial – the proof of consistency of combinatory logic shows that S and K belong to distinct congruence classes.

Term–Model
T = terms from combinatory logic
$t \sim t'$ (equivalently $t = t'$ provable). Let $\bar{t} = \{t' : t \sim t'\}$
Notice that $t_1 \sim t_1'$, $t_2 \sim t_2'$ implies $t_1 * t_2 \sim t_1' * t_2'$ (congruence).
For $D = \{\bar{t} : t \in T\}$ with $\bar{t}_1 * \bar{t}_2 = \overline{t_1 * t_2}$,
the structure $\mathcal{D} = \langle D, *, S, K \rangle$ is a combinatory
algebra – the term–model.

The reader may like to reflect on the analogy between this construction and the construction of the field of rational numbers from the natural numbers. There we form equivalence classes of triples $\langle a, b, c \rangle \in \mathbb{N}^3$ by setting $\langle a, b, c \rangle \sim \langle a', b', c' \rangle$ if the equation $ac' + b'c = a'c + bc'$ can be proved. But here there lies a qualitative difference: the numerical calculation, which leads to a proof (or to a disproof) of the given equation, is quickly carried out. And hence we are ready to describe $\mathbb{Q}$ so constructed as concrete. On the other hand, the question whether an equation $t = t'$ can be proved in combinatory logic, is incomparably harder, indeed as we shall see, algorithmically insolvable. The elements $\bar{t}$, of which the term–model consists, are not algorithmically delimitable as sets of terms. {Question: from this point of view, how does the construction of the term–model compare with that of the real numbers using Dedekind sections?}

The linkage of the construction of models to concept of proof is somewhat unusual for the customary set theoretic presentation of mathematical structures, to put it mildly. And it was only forty years later that the first set–theoretic constructions were known – those of Scott and Plotkin (compare Barendregt). Since then a simplified, and as we will see, universal model construction has come about. This is presented in what follows.

We seek to realize the elements of combinatory algebra as sets. For functions f from sets to sets F the graph of f is again a set, namely a set of pairs $\langle a, f(a)\rangle$. Application of the function f, understood as graph, to an argument a, understood as $\{a\}$, takes the form of an operation on sets: $F\triangle\{a\} = \{b : \exists x \in \{a\}.\langle x, b\rangle \in F\}$. This operation $\triangle$ is defined on arbitrary sets M and N of set theory: $M\triangle N = \{y : \exists x \in N.\langle x, y\rangle \in M\}$. However, this operation will not make a combinatory algebra from set theory, but the following will:

$$M*N := \{y : \exists x \subseteq N. \langle x, y\rangle \in M \wedge x \text{ finite }\}.$$

Yet it seems rather excessive to exploit set theory as a whole for this construction, and it seems more appropriate only to construct those sets and sets of sets etc., that will later really be needed in the proof of combinatory completeness. This consideration leads to the following construction. In order to emphasize the functional connection we write $(x \to y)$ instead of $\langle x, y\rangle$, and moreover, in future, we us lower case greek letters "α", "β" ... to denote finite subsets.

Graph Models

Let $A \neq \emptyset$ and $G_n(A)$ be recursively defined by $G_0(A) = A$,
$G_{n+1}(A) = G_n(A)\cup\{(\alpha \to y) : \alpha \subseteq G_n(A), \alpha \text{ finite}, y \in G_n(A)\}$.
Let $G(A) = \bigcup\limits_{n\geq 0} G_n(A)$ and let D_A be the power set of $G(A)$.
Then $\mathcal{D}_A = \langle D_A, *, S, K\rangle$ is a combinatory algebra where

$$
\begin{aligned}
M * N &= \{y : \exists\alpha \subseteq N.(\alpha \to y) \in M\}, \\
K &= \{(\{y\} \to (\emptyset \to y)) : y \in G(A)\}, \\
S &= \{((\tau \to (\{r_1, \ldots, r_n\} \to s)) \\
&\quad \to (\{\sigma_1 \to r_1, \ldots, \sigma_n \to r_n\} \to (\sigma \to s)) : \\
&\quad n \geq 0, r_1, \ldots, r_n \in D_A, \tau \cup \bigcup \sigma_i = \sigma \subseteq G_A, \sigma \text{ finite}\}.
\end{aligned}
$$

Here we must carry out two things, first show that $\mathcal{D}_A$ is really a combinatory algebra, and secondly that our construction has the promised universal character. The latter follows from the following theorem.

Embedding Theorem *Let $\mathcal{A} = \langle A, \cdot\rangle$ be an algebraic structure with binary operation $\cdot$. Then $\mathcal{A}$ can be isomorphically embedded in $\mathcal{D}_A = \langle D_A, *\rangle$.*

Proof. The embedding $f : A \to D_A$ is put together from maps $f_i : A \to D_A$, with $f(a) = \bigcup\limits_{n\geq 0} f_i(a)$ where for $a \in A$:

$$f_0(a) = \{a\}$$
$$f_{n+1}(a) = f_n(a) \cup \{(\{a'\} \to b) : a' \in A, b \in f_n(a \cdot a')\}.$$

Since $f(a) \cap A = \{a\}$, f is injective. We also have

$$f(a) * f(b) \;=\; f(a \cdot b) \;:$$

$$
\begin{aligned}
f(a) * f(b) &= \{y : \exists \alpha \subseteq f(b).\,(\alpha \to y) \in f(a)\} \\
&= \textstyle\bigcup_{0 \le i}\{y : \exists \alpha \subseteq f(b).\,(\alpha \to y) \in f_{i+1}(a)\} \\
&= \textstyle\bigcup_{0 \le i}\{y : (\{b\} \to y) \in f_{i+1}(a)\} \\
&= \textstyle\bigcup_{0 \le i}\{y : y \in f_i(a \cdot b)\} = f(a \cdot b). \quad \square
\end{aligned}
$$

For the proof that $\mathcal{D}_A$ is a combinatory algebra one uses rather mechanical verification. Thus let M, N, L be arbitrary subsets of $G(A)$; we have:

$$
\begin{aligned}
KMN &= \{s : \exists \alpha \subseteq N \,\exists \beta \subseteq M.\,(\beta \to (\alpha \to s)) \in K\} \\
&= \{s : \{s\} \subseteq M.\,(\{s\} \to (\emptyset \to s)) \in K\} = M\,. \\
ML(NL) &= \{s : \exists \rho \subseteq NL.\,(\rho \to s) \in ML\} \\
&= \{s : \exists n \ge 0 \,\exists r_1,\ldots,r_n \in G(A) \exists \sigma_1,\ldots,\sigma_n \subseteq L. \\
&\qquad (\{r_1,\ldots,r_n\} \to s) \in ML \wedge (\sigma_1 \to r_1),\ldots,(\sigma_n \to r_n) \in N\} \\
&= \{s : \exists n \ge 0 \,\exists r_1,\ldots,r_n \in G(A) \exists \sigma_1,\ldots,\sigma_n \subseteq L \exists \tau \subseteq L. \\
&\qquad (\{\tau \to (\{r_1,\ldots,r_n\} \to s)) \in M \wedge (\sigma_1 \to r_1),\ldots,(\sigma_n \to r_n) \in N\}\,. \\
SMNL &= \{s : \exists \sigma \subseteq L \,\exists \eta \subseteq N \,\exists \epsilon \subseteq M.\,(\epsilon \to (\eta \to (\sigma \to s))) \in S\} \\
&= \{s : \exists \sigma \subseteq L \,\exists n \ge 0 \,\exists r_1,\ldots,r_n \in G(A) \exists \tau,\sigma_1,\ldots,\sigma_n \subseteq L. \\
&\qquad (\tau \to (\{r_1,\ldots,r_n\} \to s)) \in M \wedge \\
&\qquad (\sigma_1 \to r_1),\ldots,(\sigma_n \to r_n) \in N \wedge \sigma = \tau \cup \textstyle\bigcup \sigma_i\} \\
&= \{s : \exists n \ge 0 \,\exists r_1,\ldots,r_n \in G(A) \exists \tau,\sigma_1,\ldots,\sigma_n \subseteq L. \\
&\qquad (\tau \to (\{r_1,\ldots,r_n\} \to s)) \in M \wedge (\sigma_1 \to r_1),\ldots,(\sigma_n \to r_n) \in N\}\,. \\
S \ne K &: \quad (\{a\} \to (\emptyset \to a)) \notin S\,.
\end{aligned}
$$

Therefore we have:

Existence Theorem $\mathcal{D}_A$ *is a combinatory algebra.* $\square$

As a useful illustration of combinatory completeness we consider a typical algebraic problem, namely the question of the explicit solvability of equations in a combinatory algebra, especially of equations of the form

$$F \cdot X = X\,.$$

An explicit solution of such a so–called fixed point equation arises, if a solution X can be written out as a combination of F and the basic combinators S and K. However then a description in the form YF can also be found. This is made possible by the following construction: Let D be the combinator, which exists because of combinatory completeness and satisfies the equation

$$\boldsymbol{D}xy = x(yy)\,,$$

and let $\boldsymbol{Y}$ be the combinator with the defining equation

$$\boldsymbol{Y}x = \boldsymbol{D}x(\boldsymbol{D}x)\,.$$

A fixed point of F, i.e. a solution of $FX = X$ is now given by $\boldsymbol{Y}F$. Indeed

$$\boldsymbol{Y}F = \boldsymbol{D}F(\boldsymbol{D}F) = F((\boldsymbol{D}F)(\boldsymbol{D}F)) = F(\boldsymbol{Y}F)\,.$$

Theorem $\boldsymbol{Y}F$ *is the least fixed point of F in the graph model.*

Proof.   We have just proved that $\boldsymbol{Y}F$ is a fixed point. Now let X be an arbitrary fixed point. We must show that $\boldsymbol{Y}F$ considered as a subset of $G(A)$ is also a subset of X. Let $M = \boldsymbol{D}F$, so that $\boldsymbol{Y}F = MM$.

(a) $\boldsymbol{Y}F = MM \subseteq X$ comes from the fact that $\alpha\alpha \subseteq X$ for all finite $\alpha \subseteq M$. Thus: if $a \in MM$, there exists $\alpha \subseteq M$ with $(\alpha \to a) \in M$. But then there exists some finite β, for example $\alpha \cup \{\alpha \to a\}$, with $a \in \beta\beta$.

(b)   Let α be an arbitrary finite subset of M. Since $M \subseteq G(A) = \bigcup_n G_n(A)$, then each finite α is a subset of $G_n(A)$ for some smallest n. We can therefore set up an inductive proof. If $\alpha \subseteq G_o(A) = A$, then $\alpha \cdot \alpha = \emptyset \subseteq X$. Let $\alpha\alpha \subseteq X$ for all $\alpha \subseteq M$, $\alpha \subseteq G_{n-1}(A)$, and let $\beta \subseteq G_n(A), \beta \not\subseteq G_{n-1}(A)$, $\beta \subseteq M$, $a \in \beta\beta$. We must show that $a \in X$. Since $a \in \beta\beta$, there exists some $\gamma \subseteq \beta$ with $(\gamma \to a) \in \beta$. From this it follows that $\gamma \subseteq G_{n-1}(A)$ and therefore by the inductive assumption, that $\gamma\gamma \subseteq X$. Since $(\gamma \to a) \in \beta \subseteq M$, clearly $a \in M\gamma = F(\gamma\gamma)$ by definition of M. But it is now true in general that from $p \subseteq q \subseteq G(A)$ it follows that $Fp \subseteq Fq$. Since $\gamma\gamma \subseteq X$, it is in particular true that $F(\gamma\gamma) \subseteq FX = X$, therefore that $a \in F(\gamma\gamma) \subseteq X$ and $a \in X$. $\square$

In the graph model the least fixed point is not only, as above, combinatorially – explicitly definable, but also in a very appealing way set–theoretic. Namely, as is easy to show, it is

$$\bigcup_n F^n\emptyset\,,$$

where $F^o = \boldsymbol{I}, F^{n+1} = \boldsymbol{B}F^nF$, i.e. therefore $F^o\emptyset = \emptyset$, $F^{n+1}\emptyset = F^n(F\emptyset)$.

Further Reading

Barendregt, H.P.: The type free lambda calculus, in: J. Barwise: Handbook of Mathematical Logic. pp. 1091-1132, Amsterdam, North-Holland, 1977

Engeler, E.: Algebras and Combinators, Algebra Universalis 13, pp. 389-392, (1981)

Meyer, A.: What is a model of the Lambda Calculus?, Report MIT/LCS/TM-171 (resp. TM-201), 1980 (1981)

Schellinx, H.: Isomorphismus and nonisomorphisms of graph models, J. of Symbolic Logic, 56, pp. 227–249, (1991)

Longo, G.: Set-Theoretical Models of λ–Calculus: Theories, Expansions, Isomorphism, Annals of Pure and Applied Logic, 24, pp. 153-188, (1983)

§ 4 Lambda Calculus

Let us remind ourselves once again what "combinatory algebra" means: for each term $t(x_1, \ldots, x_n)$ there exists an element T in D_A, so that for arbitrary $M_1, \ldots, M_n \in D_A$ we have

$$TM_1 M_2 \ldots M_n = t(M_1, \ldots, M_n);$$

the algorithmic rule becomes a concrete object. It lies in the nature of the subject, that by applications we always pass from a term t to an object T. Something is uncomfortable here – there may be many different T for one t – here is a simple example. Above we have used a combinator I with the property $I * N = N$ for all N. In D_A there exist at least two such I:

$$I_0 = \{(\{a\} \to a) : a \in G(A)\} \text{ and}$$
$$I_1 = \{(\alpha \to a) : a \in \alpha \subseteq G(A)\},$$

and indeed many, infinitely many, more (which?). The way out, when passable, consists in giving a construction which to each M associates a unique $\overline{M}$ in the given combinatory algebra, which for each N behaves well under application: $MN = \overline{M}N$. Of course we again want to realize such a construction itself as a combinator L. This leads to the definition of combinatory models.

Combinatory Models

A combinatory model is a structure

$$\mathcal{D} = \langle D, *, S, K, L \rangle$$

with a binary operation $*$ and elements S, K, L which satisfy the following requirements:

$K xy = x$

$S xyz = xz(yz)$

$L xy = xy$

$(\forall y . x_1 y = x_2 y) \supset L x_1 = L x_2$

In the graph model L is to be defined as a subset of $G(A)$:

Lemma *With $L = \{\{\alpha \to b\} \to (\beta \to b) : \alpha \subseteq \beta \subseteq G(A) , b \in G(A)\}$, for finite α, β, the graph model yields a combinatory model.*

Proof. If $b \in MN$, then there exists $\alpha \subseteq N$ with $(\alpha \to b) \in M$; then $\alpha \to b \in LM$ and thus $b \in LMN$. Conversely, if $b \in LMN$, then $(\beta \to b) \in LM$ for some $\beta \subseteq N$, hence $(\alpha \to b) \in M$ for some $\alpha \subseteq \beta$ and therefore $b \in MN$. It follows that $LMN = MN$ for all M and N.

For the proof of the second requirement we assume that $LM_1 \neq LM_2$, thus for example $(\alpha \to b) \in LM_1$ and $(\alpha \to b) \notin LM_2$. From $(\alpha \to b) \in LM_1$ follows the existence of some $\beta \subseteq \alpha$ with $(\beta \to b) \in M_1$. For this $\beta, b \in M_1\beta$. Since on the other hand $(\alpha \to b) \notin LM_2$, there is no $\gamma \subseteq \alpha$ with $(\gamma \to \beta) \in M_2$, in particular $(\beta \to b) \notin M_2$ and so $b \notin M_2\beta$. Hence it is not true that $M_1 N = M_2 N$ for all N. □

In combinatory models it is possible, as we have seen, to associate a uniquely determined concrete element LT, with each term $t(x)$ for which $LTx = t(x)$ for all x. T is constructed as an expression in S and K using combinatory completeness. Now this manner of presentation, in particular its decription, is at the same time both round–about and so central, that it forces the introduction of a precise and pregnant symbolism for this functional abstraction. In contrast to set theoretic abstraction, the passage from a property $E(x)$ to its extension $\{x : E(x)\}$, function abstraction has not yet made its way into contemporary mathematics, even though it might contribute much to the highlighting of its concepts.

In order to illustrate the *Lambda Notation*, whose formal introduction follows, we want to exhibit two examples from daily mathematical practice. What actually is the function $3x^2 + 1$? If one wants to be exact, one usually introduces a function sign, say f, and says: "The function $f : \mathbb{R} \to \mathbb{R}$ defined by $f(x) = 3x^2 + 1$". Here x is clearly a variable, which in this connection can be relabelled as another y without change in meaning; it is treated as a "bound variable".

The lambda–notation removes the arbitrariness of the choice of "f" as function symbol; for f it offers the expression "$\lambda x. 3x^2+1$". Thus $(\lambda x. 3x^2+1)2$ is to be evaluated to 13. In this way the Lambda–notation overcomes the *symbolic insufficiency* of the usual language of analysis.

Let P be a functional, i.e. the association of functions to functions. In the language of the naive mathematician one can for example define: "Let $f(x)$ be a function and let $g(x) = P(f(x))$." This is generally open to *misunderstanding*; for example, what is $P(f(x-1))$? Is it $P(f)(x-1)$ or $P(g)(x)$, where $g(x) = f(x-1)$? These are not the same, as the following example shows:

Let P be defined by

$$P(f)(x) = \begin{cases} 0 \text{ for } x \leq 0, \\ f(x) \text{ otherwise} . \end{cases}$$

Then

$$P(f)(x - 1) = \begin{cases} 0 \text{ for } x \leq 1, \\ f(x - 1) \text{ otherwise} ; \end{cases} \qquad P(g)(x) = \begin{cases} 0 \text{ for } x \leq 0, \\ f(x - 1) \text{ otherwise} . \end{cases}$$

The mathematician undrilled in formalism in such cases habitually introduces symbols and conventions, which set such misunderstandings aside; the introduction of differential operators is a case in point! Again the lambda-notation offers a uniform and elegant method for the same end.

Here we consider one and only one application of the lambda–notation – namely as notation for the term LT, which was constructed for $t(x)$ so as to guarantee $LTx = t(x)$ for all x. We will denote this term then by $\lambda x . t(x)$. More precisely:

Definition *Let $t(x_1, \ldots, x_n, x)$ be a term in combinatory logic, built up from S, K, L and the variables $x_1, \ldots, x_n, x$. Let T be the term in S, K, L and $x_1, \ldots, x_n$, which is constructed following the lemma of § 1, and for which we have: $T \cdot x = t(x_1, \ldots, x_n, x)$ for all $x_1, \ldots, x_n$. Then $\lambda x . t(x_1, \ldots, x_n, x)$ denotes the term LT.*

Notation is nothing without rules for its use; in the present case these take the developed form of a calculus. Historically this calculus was created by Church independently of combinatory logic, and indeed first of all as a purely formal calculus; i.e. without prior formulation of questions concerning the concretization of elements in some or other, perhaps set–theoretic, frame. The reversal of the sequence: combinatory algebra before lambda calculus is carried out here for expository reasons. (Compare the introduction to § 2.)

Convention For terms t_1, t_2 and variables x let $t_1|_x^{t_2}$ be the result of replacing x in all occurrences in t_1 by t_2. By appropriate renaming we also avoid putting a variable in t_2 into the scope of some λ in the course of a substitution.

Lambda Conversion Calculus

Language:
Terms built up from atomic terms, i.e. from variables $x, y, z, \ldots$, (and optionally some constants) with the help of the binary operation $*$ and lambda abstraction: If t is a term and x a variable, then $(\lambda x. t)$ is again a term.
Formulae: equations between terms.

Axioms:
$t = t$ for atomic terms
$(\lambda x. t) = (\lambda y. t|_x^y)$ (renaming)
$(\lambda x. t_1)t_2 = t_1 |_x^{t_2}$ (β– reduction)

Deductive Rules:

$$\frac{t_1 = t_2}{t_1 t_3 = t_2 t_3} \quad \frac{t_1 = t_2}{t_3 t_1 = t_3 t_2} \quad \frac{t_1 = t_2}{t_2 = t_1}$$

$$\frac{t_1 = t_2 \; t_2 = t_3}{t_1 = t_3} \qquad \frac{t_1 = t_2}{\lambda x. t_1 = \lambda x. t_2}$$

Provability:
As in combinatory logic.

Theorem *Each combinatory model is a model of the lambda conversion calculus; i.e. one can so define the operation of lambda abstraction in combinatory models, such that the axioms hold and the validity of equations under the deductive rules is maintained.*

Proof.   We have set up everything in the best possible way. Suppose that $\mathcal{D} = \langle D, *, S, K, L \rangle$ is a combinatory model. The validity of an equation $t_1 = t_2$ in $\mathcal{D}$ means the assertion that for each assignment of values from D to the variables the result is the same for the terms t_1 and t_2. By "result" we mean here – following the construction of terms, the results of applying the application operation $*$ or the lambda abstraction operator, as it is given above in the definition of this operation (by LT). With this hint the proof follows by direct verification. Thus the proof of the β –reduction axiom consists of the following induction of the construction of t_1.

If t_1 is the variable x, then by definition T is fixed as SKK, and we have $\lambda x. t_1 = LT = L(SKK)$. Hence $(\lambda x. t_1)t_2 = L(SKK)t_2 = t_2 = x|_x^{t_2}$. If t_1 is another variable or S, K or L, then T is defined as Kt_1, and we have $(\lambda x. t_1)t_2 = L(Kt_1)t_2 = Kt_1 t_2 = t_1 = t_1|_x^{t_2}$. Finally, if the term is composite, say $t_1 * t_2$, then the corresponding T is defined by ST_1T_2, where T_1 and T_2 can be found by the inductive assumption. With this we have

$$
\begin{aligned}
(\lambda x.\, t_1 t_2) t_3 &= \boldsymbol{L}(\boldsymbol{S} T_1 T_2) t_3 = \boldsymbol{S} T_1 T_2 t_3 = T_1 t_3 (T_2 t_3) \\
&= \boldsymbol{L} T_1 t_3 (\boldsymbol{L} T_2 t_3) = t_1 |_x^{t_3} * t_2 |_x^{t_3} = (t_1 * t_2) |_x^{t_3} \ .
\end{aligned}
$$

The last deductive rule is, for example, proved as follows: Let $t_1 = t_2$, i.e. $t_1(x) = t_2(x)$ is satisfied for all choices of x from among the elements of D. Again suppose T_1 and T_2 so determined, that $t_1(x) = T_1 * x$, $t_2(x) = T_2 * x$. Then $T_1 * x = T_2 * x$ for all x, and in our combinatory model, $\boldsymbol{L} T_1 x = \boldsymbol{L} T_2 x$ for all x in D. By definition of the notation, we therefore have $\lambda x.\, t_1 = \lambda x.\, t_2$ in $\mathcal{D}$. □

It may have occurred to the reader that in the lambda conversion calculus we may have allowed constants such as $\boldsymbol{S}, \boldsymbol{K}, \boldsymbol{L}$, but that these are not necessary in the setting up of the calculus. Indeed elements with the appropriate combinatory properties can be easily defined by means of λ–terms, for example $\boldsymbol{S}$ is represented by $\lambda x.\lambda y.\lambda z.\, xz(yz)$. Furthermore it may be considered that the definition of combinatory models is spoiled by the form of its additional axiom, an implication. Indeed more recent investigations have given an axiomatization, which only consists of equations (compare Barendregt, Chap. 7).

Further Reading

Barendregt, H.P.: The Lambda Calculus, its Syntax and Semantics, Studies in Logic 103, Amsterdam, North–Holland, 1981

Church, A.: The Calculi of Lambda–Conversion, Princeton, NJ, Princeton University Press, 1941

Scott, D.pp.: Lambda calculus: some models, some philosophy, In: Barwise et al.: The Kleene Symposium, Studies in Logic 101, pp. 381-421, Amsterdam, North–Holland, 1980

Scott, D.S.: Relating theories of the λ–calculus, in: Seldin et al.: To H.B. Curry; Essays on Combinatory Logic, Lambda Calculus and Formalism, pp. 403–450, New York, Academic Press, 1980

§5 Computability and Combinators

In the present section we want briefly to go into the connections between combinatory algebra and logic, and recursion theory. This will serve as a demonstration that concept formation in combinatory algebra has achieved its declared aim. We started out from the idea of capturing "algorithmic rules" by objects in an algebraic structure. But this is known to be also accomplished by the notion of a partially–recursive function; this is Church's thesis. It therefore remains to show that each partially–recursive function corresponds to a combinator, which, applied to suitable numerical objects, does the same job.

First we make use of the possibility of building numerical objects explicitly from the start into the graph model $\mathcal{D}_A$, by choosing the initial set A to be $\mathbb{N}$. We can associate each natural number n with an element $\{n\}$ of $\mathcal{D}_A$ as representative; we abbreviate $\{n\}$ as $\underline{n}$. Furthermore we want to choose the basic functions of recursion theory, namely the

$$\text{Successor function} \quad N : N(n) = n+1, \quad n \in \mathbb{N} \quad \text{and}$$
$$\text{Zero function} \quad Z : Z(n) = 0, \quad n \in \mathbb{N},$$

as specific elements of $G(\mathbb{N})$, namely

$$\boldsymbol{N} = \{\{n\} \to n+1 : n \in \mathbb{N}\} \text{ and}$$
$$\boldsymbol{Z} = \{\{n\} \to 0 : n \in \mathbb{N}\}.$$

Projection functions also belong to the basic functions of recursion theory

$$U_i^m : U_i^m(n_1, \dots, n_n) = n_i, \; 1 \leq i \leq m \in \mathbb{N}.$$

For U_i^m we have a general combinator ready to use with defining equation

$$\boldsymbol{U}_i^m x_1 x_2, \dots, x_m = x_i.$$

The objects so chosen from the combinatory algebra $\mathcal{D}_A$ closely mirror the corresponding number functions: we have

$$\boldsymbol{N}\underline{n} = \underline{m} \qquad \text{iff} \quad N(n) = m,$$
$$\boldsymbol{U}_i^m \underline{n_1}\underline{n_2} \dots \underline{n_m} = \underline{k} \quad \text{iff } U_i^m(n_1, \dots, n_m) = k,$$

and similarly for $\boldsymbol{Z}$. In general:

Definition *A partial function* $f : \mathbb{N}^n \to \mathbb{N}$ *is called* representable *in* $\mathcal{D}_A$, *if there exists an object* $\boldsymbol{F}$ *in* $\mathcal{D}_A$, *so that*

$$\boldsymbol{F}\underline{k_1}\underline{k_2} \dots \underline{k_n} = \underline{m} \qquad \text{holds in } \mathcal{D}_A \text{ iff}$$
$$f(k_1, k_2, \dots, k_n) \qquad \text{is defined and equals } m.$$

Theorem *Each partially–recursive function is representable in the graph model (as enlarged by N, Z and U_i^m).*

Proof. As definition of a partially recursive function we use the one which uses the zero function Z, the successor function N and the projection functions U_i^m as basic functions, and closes these with the schemes of composition, primitive recursion and the μ–scheme.

(a) The representability of the basic functions is built into the construction of $\mathcal{D}_A$ just as are the number objects.

(b) Closure under composition:
Let $f(x_1,\dots,x_n) = h(g_1(x_1,\dots,x_n),\dots,g_m(x_1,\dots,x_n))$, where h is represented by H and each g_i by G_i. Because of combinatory completeness there exists F with $Fx_1\dots x_n = H(G_1x_1\dots x_n)\dots(G_mx_1\dots x_n)$. We verify the representation: $F\underline{k}_1\dots\underline{k}_n = H(G_1\underline{k}_1\dots\underline{k}_n)\dots$ $\dots(G_m\underline{k}_1\dots\underline{k}_n)$. When all $g_i(\underline{k}_i\dots\underline{k}_n)$ are defined this equals

$$H\underline{g_1(k_1,\dots,k_n)}\dots\underline{g_m(k_1,\dots,k_n)} = \underline{h(g_1(k_1,\dots,k_n),\dots,g_m(k_1,\dots,k_n))}$$

under the corresponding condition.

(c) Closure under primitive recursion:
For the sake of clarity we restrict ourselves to the case of single variable functions; suppose therefore that

$$f(n) = \text{if } n = 0 \text{ then } k \text{ else } g(f(n-1),n-1),$$

where g is represented by G. We consider applying the fixed point combinator Y, but this needs some preparation. For this first of all we need a predecessor function, but only for numbers $n \geq 1$, and a decision function, which corresponds to "if...then...else". We write

$$V = \{\{n+1\} \to n : n \in \mathbb{N}\},$$

where clearly $V\underline{n} = \underline{k}$ iff $n > 0$ and $k = n-1$. For the decision function we need an object $Zero$ with

$$Zero\ xMN = \begin{cases} M & \text{if} & x = \underline{0} \\ N & \text{if} & x = \underline{n+1} \end{cases} \quad \text{for some } n \in \mathbb{N},$$

(for arbitrary objects M, N). The dependence of $Zero$ on the non–number object x is irrelevant. As an object in $\mathcal{D}_A$ $Zero$ can be defined by

$$Zero := \{\{0\} \to x : x \in K\} \cup \{\{n+1\} \to x : x \in KI, n \in \mathbb{N}\}.$$

We then have $Zero\ \underline{0} = K$ and $Zero\ \underline{n+1} = KI$, so that $Zero\ \underline{0}MN = KMN = M$, and $Zero\ \underline{n+1}\ MN = KIMN = IN = N$. Using the combinators of V and $Zero$ we can now write down primitive recursion as a fixed point equation. For this suppose, according to combinatory completeness, that

$$\boldsymbol{R}fx = \boldsymbol{Zero}\ x\underline{k}(G(f(\boldsymbol{V}x))\boldsymbol{V}x)\,.$$

The desired representative for f is the fixed point of

$$\boldsymbol{R}f = f\,,$$

given by $\boldsymbol{YR}$. Then $\boldsymbol{YR}$ does indeed solve the recursion equations, namely:

$$
\begin{aligned}
\boldsymbol{YR}\underline{0} \ &= \ \boldsymbol{R}(\boldsymbol{YR})\underline{0} = \boldsymbol{Zero}\ \underline{0}\,\underline{k}(G((\boldsymbol{YR})(\boldsymbol{V}\underline{0}))(\boldsymbol{V}\underline{0})) = \underline{k} = \underline{f(0)}\,,\\
\boldsymbol{YR}\underline{n}+1 \ &= \ \boldsymbol{R}(\boldsymbol{YR})\underline{n}+1 \\
&= \ \boldsymbol{Zero}\ \underline{n}+1\underline{k}(G((\boldsymbol{YR})(\boldsymbol{V}\underline{n}+1))(\boldsymbol{V}\underline{n}+1))\\
&= \ G((\boldsymbol{YR})\underline{n})\underline{n} = G\underline{f(n)}\,\underline{n} = \underline{g(f(n),n)} = \underline{f(n+1)}\,.
\end{aligned}
$$

Since $\boldsymbol{YR}$ is at the same time the smallest fixed point, it is guaranteed that it also represents the intended solution of the primitive recursion equation.

(d)   Closure with respect to the μ- scheme.
Let $f(x) = \mu y[g(x,y) = 0]$ (i.e. $f(x)$ is the least y, if it exists, for which $g(x,y)$ is defined and $= 0$) and let g be represented by $\boldsymbol{G}$. Now f can be introduced as follows with the aid of if ... then ... else:

$$h(x,y) = \textbf{if}\ g(x,y) = 0\ \textbf{then}\ y\ \textbf{else}\ h(x,y+1);\ f(x) = h(x,0)\,.$$

Correspondingly we consider

$$\boldsymbol{R}hxy = \boldsymbol{Zero}\ (Gxy)y(Hx(Ny))\,,\boldsymbol{H} = \boldsymbol{YR}\,,\boldsymbol{F}x = Hx\underline{0}\,.$$

For the verification we calculate as follows:

$$
\begin{aligned}
\boldsymbol{F}\underline{n} \ &= \ H\underline{n}\,\underline{0} = (\boldsymbol{YR})\underline{n}\,\underline{0} = \boldsymbol{R}(\boldsymbol{YR})\underline{n}\,\underline{0} = \boldsymbol{RH}\underline{n}\,\underline{0}\\
&= \ \boldsymbol{Zero}\ (G\underline{n}\,\underline{0})\underline{0}(H\underline{n}\,\underline{1}) = \left\{
\begin{array}{ll}
\underline{0}\,, & \text{if}\ g(n,0) = 0\,,\\
H\underline{n}\,\underline{1} & \text{otherwise}\,.
\end{array}
\right.
\end{aligned}
$$

For $H\underline{n}\,\underline{1}$ we have the values

$$H\underline{n}\,\underline{1} = \boldsymbol{RH}\underline{n}\,\underline{1} = \boldsymbol{Zero}\ (G\underline{n}\,\underline{1})\underline{1}(H\underline{n}\,\underline{2}) = \left\{
\begin{array}{ll}
\underline{1}\,, & \text{if}\ g(n,1) = 0\,,\\
H\underline{n}\,\underline{2} & \text{otherwise}\,.
\end{array}
\right.$$

And so on for $H\underline{n}\,\underline{2}, H\underline{n}\,\underline{3},\dots$. In this way and in each case the correct value of $\boldsymbol{F}\underline{n}$ is produced. No value results, as indeed is our wish, if any one of the $G\underline{n}\,\underline{i}$ to be evaluated gives no numerical value, for this case $f(n)$ is also undefined.   □

In this way we have shown, that in *certain* combinatory algebras all computable functions can be represented by objects. However, in reality this is possible for all combinatory algebras. In order to prove this fact it suffices to put number objects $\mathbf{0,1,2},\ldots$ in arbitrary combinatory algebras and to replace the ad hoc constructions for $\boldsymbol{N}, \boldsymbol{Z}, \boldsymbol{V}$ and $\boldsymbol{Zero}$ by general combinators.

However here we combine this step into the general with the switch to a formal standpoint: natural numbers n will be introduced as specified combinators $\boldsymbol{n}$ in the language of combinatory logic. Our knowledge is then restricted to those formulae, equations, which are derivable in combinatory logic. Correspondingly the concept of representability is to be reformulated as

---

Definable Number Functions

A partial function $f : \mathbb{N}^n \to \mathbb{N}$ is called definable, if there exists a term $\boldsymbol{F}$ in combinatory logic so that

$$\boldsymbol{F k_1 \ldots k_n = m} \text{ is provable iff}$$

$$f(k_1, \ldots, k_n) \text{ is defined and equal to } m.$$

Now we have to carry out the generalizations indicated above in a formal framework, hence to define the combinators $\boldsymbol{n}(n \in \mathbb{N}), \boldsymbol{N}, \boldsymbol{Z}, \boldsymbol{V}$ and $\boldsymbol{Zero}$. Then we have the

Theorem *Each partially recursive function is definable (in combinatory logic).*

In order to simplify the following definition we want to make use of the lambda notation for particular combinators: denote by $\lambda x. t(x)$ the term T constructed from $\boldsymbol{S}$ and $\boldsymbol{K}$ in §1. Then $Tx = t(x)$ for all x. In order to economize on brackets we make the convention that the symbol

$$\lambda xyz. t \text{ stands for } \lambda x. \lambda y. \lambda z. t,$$

for which

$$(\lambda xyz. t(x,y,z))MNL = t(M,N,L),$$

and analogously for every other number of bound variables.

The Ordered Pair

$\boldsymbol{P}$	$:= \lambda uv. \lambda x. xuv$	(Formation of pairs)
$\boldsymbol{P_0}$	$:= \lambda u. u\boldsymbol{K}$	(First component)
$\boldsymbol{P_1}$	$:= \lambda u. u(\boldsymbol{K I})$	(Second component)

Abbreviations: $\boldsymbol{P}uv = [u,v], \boldsymbol{P_0}u = (u)_0, \boldsymbol{P_1}(u) = (u)_1$

Lemma *The constructs P, P_0, P_1 constitute the formation of a pair, thus*

$$\lfloor u, v \rfloor_0 = u, \qquad \lfloor u, v \rfloor_1 = v.$$

Proof. By manipulation

$$[u, v]_0 = P_0(Puv) = (\lambda x.\, xuv)K = Kuv = u,$$
$$[u, v]_1 = P_1(Puv) = (\lambda x.\, xuv)(K\,I) = (K\,I)uv = Iv = v.$$

Natural Numbers

Recursively define combinators $0, 1, 2 \ldots$, which define natural numbers:

$$0 := I, 1 := [0, K], \ldots, n+1 := [n, K].$$

Are the combinators associated to the different natural numbers themselves different? From the formal point of view difference can only mean that $n \neq m$ can be proved. This however has no sense, since inequality does not belong to the formulae of combinatory logic, still less to the provable formulae. In § 2 under comparable circumstances (is $S = K$?) we have taken the position, that we have just the *unprovability* of the equation to prove. In § 3 for the construction of the term model this was raised to a principle by combining terms, which are provably equal, into congruence classes. Essentially to show the inprovability of $S = K$ we used Lemma 1 plus Lemma 4. These imply that $m = n$ is provable, if and only if m and n can be reduced to a common z by a sequence of contractions. Since S and K certainly cannot be further reduced, $S = K$ cannot be provable.

Definition *A term t in combinatory logic is said to be in* normal form, *if it cannot be further reduced in the contraction calculus.*

Corollary *If t_1 and t_2 are syntactically distinct and in normal form, then $t_1 = t_2$ is not provable in combinatory logic.*   □

Lemma *Each combinator $n, n \in \mathbb{N}$ is in normal form.*   □

With this lemma the stage is clear, and we make the following definition:

Number Theoretic Combinators

$$Z \quad := \quad \lambda x.0$$
$$N \quad := \quad \lambda x.[x, K]$$
$$V \quad := \quad \lambda x.(x)_0$$
$$Zero \quad := \quad \lambda x.(x)_1(K\,I)K$$

The corresponding verifications are easy to carry out:

$$Zn \qquad = \quad (\lambda x.0)n = 0 .$$
$$Nn \qquad = \quad (\lambda x.[x, K])n = [n, K] = n+1$$
$$V(Nn) \qquad = \quad (\lambda x.(x)_0)(Nn) = (Nn)_0$$
$$= \quad ((\lambda x.[x, K])n)_0 = [n, K]_0 = n .$$
$$Zero\ 0 \qquad = \quad (\lambda x.(x)_1(K\,I)K)0 = (0)_1(K\,I)K$$
$$= \quad 0(K\,I)(K\,I)K = I(K\,I)(K\,I)K$$
$$= \quad (K\,I)(K\,I)K = I\,K = K .$$
$$Zero\ (Nn) \quad = \quad (\lambda x.(x)_1(K\,I)K)[n, K]$$
$$= \quad [n, K]_1(K\,I)K = K(K\,I)K = K\,I .$$

The "verifications" just presented are in reality formal proofs in combinatory logic. Inserted into the corresponding parts of the preceding theorem on the representability in the graph model, the considerations just made give the proof of the stated theorem. □

To conclude this section we want briefly to give an example for the decidability question in combinatory logic:

The Halting Problem in Combinatory Logic

Here one is concerned with the question, whether a given term t has a normal form in combinatory logic. This question is undecidable, if there exists no combinator H such that

$$Ht = \begin{cases} 0 & \text{if } t \text{ has a normal form}, \\ 1 & \text{otherwise}. \end{cases}$$

Thus if there were an algorithm which gave a decision for this question, by Church's thesis this algorithm would be realizable as a recursive function, and as such representable by a combinator H. We now want to show that no such combinator can exist.

Theorem *The halting problem in combinatory logic is undecidable.*

Proof. We assume that the combinator H exists. We use it to construct a combinator D, which is such that

$$Dt = \begin{cases} K(tt) & \text{if } (tt) \text{ has a normal form}, \\ S & \text{otherwise}. \end{cases}$$

Clearly, $D := \lambda t.\mathbf{Zero}\ (H(tt))(K(tt))S$ and Dt has a normal form for each t, in particular so does DD. But DD reduces by definition to a normal form different from DD namely that of $K(DD)$, contradiction. □

Further Reading

Hermes, H.: Aufzählbarkeit, Entscheidbarkeit, Berechenbarkeit; Einführung in die Theorie der rekursiven Funktionen, Springer–Verlag, Berlin, 1961

Kleene, S.C.: λ–definability and recursiveness, Duke Math. J. 2, pp. 344-353, (1936)

Engeler, E., Läuchli, P.: Berechnungstheorie für Informatiker, B.G. Teubner, Stuttgart, 1988

Springer-Verlag
and the Environment

We at Springer-Verlag firmly believe that an international science publisher has a special obligation to the environment, and our corporate policies consistently reflect this conviction.

We also expect our business partners – paper mills, printers, packaging manufacturers, etc. – to commit themselves to using environmentally friendly materials and production processes.

The paper in this book is made from low- or no-chlorine pulp and is acid free, in conformance with international standards for paper permanency.

Printing: Druckerei Zechner, Speyer
Binding: Buchbinderei Schäffer, Grünstadt